Mathematical Ecology

Mathematical Ecology

edited by
Peter Allen (Brussels), Werner Ebeling (Berlin),
Manfred Peschel (Berlin), Peter Schuster (Vienna),
Yuri M. Svirezhev (Moscow)

Mathematical Ecology deals with mathematical models of evolution processes, in the biosphere as a unity of growth and structure-building, with spring-up of new species, their interaction and possible extinction, with the impacts of human activities on the environment and the corresponding consequences for the biosphere.

The whole field seems to be a certain amalgam of suitable system philosophy, in which real world phenomena are considered through an ecological pair of spectacles with the help of system methodology and mathematics.

All publications published herein are of interdisciplinary interest for ecology, biology, economy and technical engineering.

Chaotic Dynamics Applied to Biological Information Processing

by John S. Nicolis

Akademie-Verlag Berlin 1987

Author

Prof. Dr. John S. Nicolis
Department of Electrical Engineering
University of Patras, Greece

ISBN 3-05-500377-2

Erschienen im Akademie-Verlag Berlin, DDR-1086 Berlin, Leipziger Str. 3—4

Lizenznummer: 202 · 100/485/87
Printed in the German Democratic Republic
Gesamtherstellung: VEB Druckerei „Thomas Müntzer", 5820 Bad Langensalza
Lektor: Dr. Reinhard Höppner
LSV 1085
Bestellnummer: 763 744 7 (9067)
02000

„Music is the effort we make
to explain to ourselves how
our brains work"

Lewis Thomas
"The Medusa and the snail"

Preface

The traditional, time honoured approach to Ecological dynamics is associated with stability analysis of coupled non-linear differential equations or coupled difference equations with either discrete or continuously distributed time lags amongst the appropriate variables. These variables represent either the instantaneous population densities of the interacting species or (in a more sophisticated formalism couched after game theoretical considerations) the probabilities of moves of the individual contestants - each trying to maximize its winnings or minimize its losses. Out of such a stability analysis (which may also be couched in statistical terms, namely the formulation of a master equation for the probability density function of coexistence of several variables and a search for asymptotic stability of its solutions), issues of coexistence, dominance, extinction, evolutionary stable strategies possessing point attractors, limit cycles, quasi-periodic behaviour or chaos are supposed to emerge. However, one should keep in mind that the variables entering the above abstract equations are not just particles "bumping" against each other via energy exchanges only, but biological organisms involved in information transactions as well.

Individual species are in fact waging a "dialectic struggle" the aim of which is the simulation of the opponent, its prediction and ultimately its control. It is this aspect of information transaction between biological organisms that the present monograph aims to fill.

Regarded then under this angle, any ecological study should, in the author's opinion, also try to familiarize the reader with current dynamical models of biological pattern formation, recognition, and selection after the modern paradigm of chaotic dynamics as applied to information processing. This aspect should make the "dynamically-prone" researcher in ecology aware of the importance of Symbolic processes exchanged between the Softwares ("Minds") of the contesting individuals and also the fact that information processing by biological central nervous systems is not the same as the processing by finite state automata (computers) : the former is a fundamental ingredient in understanding strife, conflict, effective strategies (algorithms) in the area of mutual simulation between contestants, where each "fighting" individual tries to compress the environmental stimuli he receives from the opponent and to classify him accordingly. In short, while the conventional approach to Ecological dynamics is involved in the study of the Energetic, Structural ("Hardware") type of interactions, our approach puts emphasis on the Symbolic, Functional ("Software") aspect and also tries to establish connections between Hardware and Software ("Body-Mind") Ecological interactions.

Wartburg March 1986

Contents

1. Introduction

In this paper we shall defend the thesis that chaos may provide the connection between information theory and physics, the main issue being that information is a physical entity and as such it should not be treated in absence of its dynamical (hardware) substratum.

The paper accordingly will revolve around the following question: How can we explain, on the basis of known dynamical laws and processes of physics, the communication transaction taking place between two (hierarchical) systems - i.e. not only the energetic exchanges but also the information ones? How can we dynamically implement the generation, stable storage, reliable retrieval and dissipation of information? In other words: What are the physical (dynamical) agents mediating symbolic interactions?

In the first place we have to clarify what we mean when we assert that two systems "communicate". For a physicist information is admittedly an almost "metaphysical" word. Indeed, in the current parlance of the mathematical theory of communication (invented by Claude Shannon 37 years ago and not improved considerably ever since) information processing is to be considered as a sequence of (one-to-one) mappings from (the members of) a set belonging to a (multidimensional) abstract signal-space onto (the members of) another set belonging to an abstract space of a different dimensionality; the "solution" of the problem of information transfer from set I to set II or, more concretely, from place A at time τ to place B at time $\tau + \Delta\tau$ boils down essentially to a variational principle which tries to find the optimum mapping (or code) in order to compromise two conflicting requirements: speed of transfer and reliability or degree of errorless reproduction of an abstract pattern of digits sent from I and received by II: in such a formulation all physics is left out.

Such a mentality of treating communication processes serves admirably the whole group of applications associated with typical communication engineering design (for an excellent review see Schalkwijk). People grow restless, however, when the above formalism is applied to fields such as biology, cognitive psychology and in general to interactions between "self"-organized systems (for a definition see below). The reason is that in such cases complete decoupling between physical and symbolic interactions (or between hardware and software) is no more possible or meaningful and people start asking questions about the ongoings at the interface between structure and function.

In such cases information processing is associated:

a) With cascades of mapping - transcriptions and/or translations - (not necessarily one-to-one but also multi-to-few, i.e. hierarchical) from one physical entity onto another. For example, in the generation of speech a pattern residing in a sequence of nerve pulses incident upon the vocal chords is mapped onto an air sound wave. This sound wave is intercepted by a microphone which maps the pattern onto an electromagnetic wave etc.

b) With low-energy triggers giving rise to high-energy effects (when informational channels intersect effector systems) or, to put it another way, with extreme sensitivity to initial condition

There is physical justification (see e.g. Nicolis (1985)) for the hierarchical or non one-to-one mapping between "object" and "image", that is the fact that physical perception is ambiguous. There are three distinct reasons for this:

a) The loss of evanescent waves: half of the encoded information in the spherical wave scattered from the object is irretrievably lost due to the impossibility of designing receptors for both propagating and evanescent waves, into which a spherical wave is analysed.

b) The existence of fluctuations around the average number of photons carried by each normal mode or electromagnetic degree of freedom.

c) The finite aperture of the perceiving instrument: This finiteness has two results: it distorts like a spatial filter a portion of propagating components of the impinging spherical wave and, secondly, it attributes a finite number of degrees of freedom to the receiving image. The ambiguity here is due to the fact that the spherical wave which originally mediates the information transfer from the transmitter to the receiver possesses an infinite number of degrees of freedom.

At first glance it seems that there exist irreconcilable differences between Soft- and Hardware. Specifically the difficulty lies in the fact, that, at the Hardware (physical) levels, processes follow laws (non-linear dynamics) with universal validity whereas, at the Software (symbolic) levels, processes follow rules (Markov chains) whose relativity and flexibility is amply manifested in any linguistic scheme.

What is the difference between laws and rules? Can we "reduce" the second to the first? Our conjecture at this point is that the issue involves essentially different time scales: If one allows a deterministic physical law to operate long enough, the result may be chaotic. However, the time scale is quite different, say for gravitational turbulence, "cardiac" turbulence (fibrillation), hydrodynamic turbulence or symbolic turbulence (Language?) to develop.

The time scale beyond which chaos emerges is in fact depending not only on the intrinsic non-linearities and the stability properties of the system involved but also on extrinsic reasons (initial and boundary conditions): Depending on the case this time scale may be large ($\sim 10^{15}$ years) or small ($\sim$ a few minutes). Specifically there are two basic ways of producing a Markov chain out of continuous non-linear dynamics:

1/ a) Taking one strange attractor in state-space, we first perform an (arbitrary) analog-to-digital conversion by way of a Poincaré mapping, which amounts in cutting the trajectory with a hyperplane, recording the isophasic crossings and deriving a set of discrete coupled difference equations.

1/ b) We choose an (arbitrary) partition on the discrete map and then we derive a Markov chain with the number of states being the partition number and transitional probabilities determined by the local slope of the map (see pp. 62-63 for a concrete example).

2) Concerning dynamics with more than one coexisting attractor, external "noise" can act catalytically and provide a sequential passage from attractor to attractor, thereby giving rise to a Markov chain with the number of states being equal to the number of coexisting attractors and transition probabilities determined by the type of superimposed noise (the E.E.G., for example, may provide such a model of time division multiplexing among attractors: See section 7).

The above theorizing aims at bridging the (colossal) difference in "behaviour" between say 1) a physical body which, when faced with the task of moving from place A to place B, follows simply and "compulsively" Maupertuis' variational principle and 2) an animal which "chooses" the specific pathway anew every time: In this second case the animal follows a rule or algorithm which may spontaneously modify itself depending on slight changes in the initial conditions and the parameters of the situation. We claim in short that symbolic processes are chaotic manifestations of collective properties of physical systems - at a higher hierarchical level; "Decision making" and "free will" [1] are deterministic since they are harnessed by natural laws, but unpredictable since they are chaotic; the same thing could be claimed for music and language.

Be as it may, we shall argue that the most important aim of the communication transaction is not just copying but simulation: this essentially refers to a process of modelling a physical system by another one with the obvious goal of predicting and controlling it. We say that a physical system I (see Fig. 1) simulates another (II) when it is capable, out of a finitely long time series it receives, of constructing minimal length algorithms or compressed descriptions of II. These algorithms, when introduced as inputs (programs) in a finite-state machine, will allow it to furnish, as an output, the behaviour of system II for all times, future and past.

---

1) What we really mean by "free will" is the simulation of a set of alternative pathways of action and a choice-making among them.

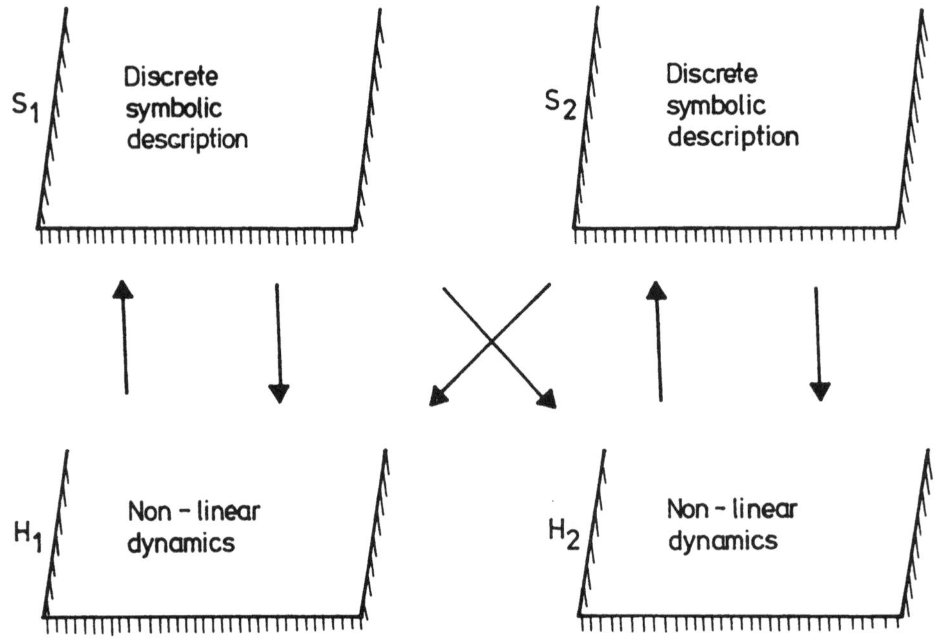

Fig. 1 Schematic layout of two communicating hierarchical systems

Since the prerequisite for simulation is the compressibility of the received time series from the partner-system (or the environment) we, in this paper, indulge in a new dynamical "paradigm", namely dissipative chaos, which through the gadgetry of "strange attractors" promises, in a single stroke, to do this: to carry out very complex functional repertoires with very simple "hardware" and to behave as a versatile (albeit somewhat uncontrollable) information processor. The new paradigm comes forth if one realizes that information is not only produced via cascading bifurcations giving rise to broken symmetry, but also via successive iterations giving rise to ever-increasing resolution. Beyond a certain resolution interval, microscopic fluctuations are no longer smeared out; they get amplified and pass from lower to higher hierarchical levels. It is these fluctuations which essentially account for new information generated by an evolving hierarchical system. (See Sections 3 and 9 for details.)

In short, the purpose is to develop an evolutionary approach for the basic problem of generation, storage and dissipation of information in physical and biological systems via dynamical processes. These processes are associated with branching-offs or cascading bifurcations giving rise to broken symmetry, as well as chaotic dissipative dynamics giving rise to decreasing and increasing resolution. From an attractor, simulating some aspects of a cognitive system (or a "processor") there exist two basic requirements: Big storage capacity and good compressibility. Conventional singularities such as stable steady states and stable limit cycles, possessing dimensionality zero and one, respectively, are very poor as information storage units but ideal as information compression devices.

Strange attractors, on the other hand, can on principle comply with both requirements via a harmonious combination of expanding and converging trajectories in state space ("positive" λ_+ and "negative" λ_- Lyapunov exponents, respectively): They may possess a considerable information dimensionality - and this fact makes them suitable for dynamical information storage; on the other hand, being "attractors", ($|\lambda_-| > \lambda_+$) they can serve as information compressors at the same time. (See the following Sections for details.)

We are going to see that (dissipative) chaotic dynamics offers an alternative and perhaps more elegant way towards the formation of hierarchy and classification: Whereas the time honoured statistical-mechanical methods lead to hierarchical structures through the collective properties of a Master equation (e.g. moments and crossmoments of the probability density function describing the dynamics at the "microscopic" level), dissipative chaotic systems accomplish the same thing in a different way: They do it by reducing the number of degrees of freedom in state space - where the dynamical system starting from a set of initial conditions, of dimensionality N sooner or later locks in at a compact subset of the state space (the "attractor") - of much lower dimensionality.

For System I to be able, in the above sense, to simulate or model System II (and vice versa), a number of sine qua non prerequisites is demanded.

In the first place, the system must be hierarchical, i.e. possess at least a hardware (energetic-structural) level and a software (cognitive symbolic, - functional) level. Information, in the form of discrete time series emanating from the S-level of System II, impinges at level H of System I, where a set of convolutions (cross-correlations) takes place among the time series received and the intrinsic dynamics of System I at level H. We will implicitly assume that this dynamics at level H is chaotic. The result of such convolutions is the emergence of collective properties possessing by far fewer degrees of freedom than S_2 or H_1. These collective properties are now transported to the level S_1.

The communication or mutual simulation between the two systems may go on, therefore, as the evolution in time of a set of pairs of coupled non-linear integral-difference equations (each associated with a given value of ξ and ρ).

$$
\left.\begin{aligned}
S_1(n+1) &= \lambda S_1(n) + \langle H_1^{\xi}(n)\; S_2^{\rho}(n)\rangle \\
S_2(n+1) &= \lambda S_2(n) + \langle H_2^{\xi}(n)\; S_1^{\rho}(n)\rangle
\end{aligned}\right\} \quad \lambda > 1, \tag{1}
$$

$H_i(n+1) = f_i(H_i(n))$ or $H_i = f_i(H)$ being the intrinsic dynamics at the hardware level(s) .

In Eqs. (1) each pair stands for one statistical moment. For each pair of the above equations the process clearly amounts to a cascading iteration on S_1 and S_2. S_1, S_2 stand for the discrete (in general multidimensional) variables at the Software level(s), H_1, H_2 stand for the (discrete or continuous) variables at the hardware levels, and n is the discrete time; ξ and ρ are positive integers. Finally $\langle\cdot\rangle$ stand for time averaging.

The harmonious co-evolution of (dynamical) bare variables and their collective properties in time is a demand for any linguistic scheme which includes semantic connotations, i.e. context and meaning: This ongoing iterative interaction (Fig. 1) between a collective property (a "set") with each of its own members and vice-versa creates the "context" and at the same time is responsible for the phenomenon of self-reference. Due to a difference in the time scales at the H and S levels we can say that the S levels may exercise history-depending, feedforward control on the H levels. Indeed, the integral component brings forth "memory" effects.

Let us divide all cognitive tasks into two broad categories:

a) "Exercises" or "deductive" problems - whose outcomes always exist , are unique and are implicitly contained in the initial premises - provided that these premises form a complete and self-consistent set, and

b) "Complex" problems, of the "Inductive" type whose outcomes may or may not exist, they may be not unique (i.e. multiple) and are not necessarily nested in the initial conditions but are in part generated by the evolution of an iterative and recursive procedure or the "flow" itself. Under such circumstances we say that if there is evolutionary (asymptotic) stability in the dynamical flow in state space, then there is a discriminable pattern or a "solution".

In dealing with problems of the category a) one goes through regressive steps on a sequence which converges, eventually "halts" and gives the solution. No new information is produced during this process although admittedly, the agent who deals with the problem may be ignorant of how to turn the "implicit" into "explicit" or how to choose amongst different strategies of approach. We bypass this issue here.

In problems of the category b), convergence cannot be guaranteed. Such problems are "dynamical" and their outcome can only be understood in a historical that is, an evolutionary context. When we say that the information required for the solution of an "inductive" problem is not "hidden" exclusively in the initial conditions - so there is no a priori truth ("α-λήθεια") to be revealed - we mean that at certain moments the evolving scenario that is the trajectory of the system in state space branches-off or bifurcates into multiple asymptotically stable regimes (stationary, periodic, quasi-periodic or aperiodic) in ways which depend on the critical values of a number of control parameters. These control parameters stand for the coupling among the variables, and their numerical values are shifted by ever-present (internal and/or external) fluctuations. When these parameters acquire their "critical" values, the coupling between the variables concerned changes in ways which precipitate instabilities - meaning that infinitesimally small changes in the parameters lead to finite changes in the variables. Beyond these instabilities the (new) structure(s) may not retain any trace of memory of the previous regimes - although it (they) evolved from them via quite deterministic processes. Similar "loss of memory" takes place in systems without bifurcations but with a sensitive dependence of their flows on initial conditions.

It is exactly this mixture of determinism and (long - or short-term) unpredictability which makes the problems of "inductive" type so relevant to information sciences.

Most systems relevant to information sciences (e.g. linguistics) contain multiple, hierarchical (see below) feedback-loops - which inevitably make them "Self"-referential. If such systems do not possess "closure" or logical consistency they are "paradoxical". The "truth(s)" - if any - revealed by such systems is associated with "Self-consistency" and this characteristic, as we will see, implies the existence of stable attractors in the iterative and recursive system, which we call "self-referential sentence". The search for self-consistency of a self-referential sentence then leads to the formulation of a boundary value problem. If so, the suggestion follows immediately that "logic" or "rationality", so far a static concept, can be dynamically implemented and replaced by "evolutionary stability".

Let us now deliberate on the concepts of hierarchy, complexity and organization, which are repeatedly used in this review.

A. Hierarchy

By a hierarchical system we mean, in general, an ensemble of interacting parts that is composed of (and is analysable or decomposable into) successively nested sets of interacting sub-units.

Each set of interacting components - forming a distinct hierarchical level - affords a specific state-space description with variables and parameters pertaining to that particular level. The interacting variables (and/or parameters) at a higher hierarchical level are "collective properties" (statistical moments or convolutions) of the dynamics going on at the level just underneath. Therefore, as we move to a higher level, we (usually but not always - see last section) witness a tremendous reduction in the number of degrees of freedom. The higher level receives selective information from "below" and in turn it exercises, on the dynamics of the lower level (efferent), feedforward control commands - which harness the further dynamical deliberations on that lower level.

Although two systems do not have to be hierarchically predisposed in order to exchange energy, in order to enter a communication transaction they certainly are so. (We need one hierarchical level to perform a cross correlation and another higher one in order to "cognize" it.)

B. Complexity

A system may be complex either on a structural or a functional level. Structural complexity increases with the number of interacting sub-units; the percentage of mutual connectedness among them, in dyadic or plural fashion; and the variance of the probability density function of the strengths of interactions between the individual sub-units.

On a functional level complexity increases with the minimum length of the (most compressed) algorithm from which one can retrieve the full behaviour of the system. In this sense, as we are going to see in the paper, the functional complexity of a system increases with the length of its "phylogenetics tail", namely its evolutionary history marked by a discrete sequence of bifurcation episodes. In that sense, then, a virus or an orange is a more complex system, functionally speaking, than a big star, since in order to understand the function to each one of them you already need as a prerequisite the existence of a supernova. But let us begin from the beginning:

Consider a macroscopic physical system at its lowest hierarchical level, namely the level of equilibrium thermal (molecular) <u>mixing.</u> At that particular state the system possesses the maximum number of degrees of freedom or the fullest possible symmetry - exactly equal to the symmetry of the microscopic laws of motion of its constituent interacting parts. The symmetry is then maximum, the number of degrees of freedom likewise, and the complexity is minimum since the number of bits required to specify a system at thermodynamic equilibrium is the smallest possible: To reproduce such a system one may start from almost <u>any</u> initial condition.

Imagine now that the system comes in contact with the environment; it may again settle to equilibrium when new conditions for minimization of its free energy are met; or it may happen that, as a consequence of internal and/or external microperturbations (which are inevitable due to the enormous number of the constituent components of the system and its environment), the one and only stable steady state of equilibrium may become deleted. Beyond a certain critical distance from equlibrium (which the system franchises triggered by fluctuations) at least two new, alternative stationary configurations (containing their own symmetrical branches) emerge. Suppose the system settles in one of them by a more or less random "choice". At this metastable state the system now forms, say, a spatial static pattern (i.e. a crystal) which, although it may appear quite symmetrical, has a number of degrees of freedom far below the one corresponding to the symmetry of the laws of motion of its constituent molecules. Nevertheless, we do not yet speak of "broken symmetry" but rather of "repealed" symmetry (Anderson (1972)) since the probability of "tunneling" to the symmetrical metastable state remains finite (although it rapidly decreases with the scale of the system).

If, however, by coupling the system with an external asymmetric field, we perturb the bifurcation diagram so that the alternative mirror symmetrical configurations get repressed (Nicolis and Prigogine (1981)) we witness a genuine broken symmetry, and the asymmetrical configuration at the corresponding state may be stable for a wide range of the control parameters. We say that the complexity of the system has increased, meaning that the length of instructions or the minimum amount of bits an external observer requires for the replication of the system has increased. We can imagine a succession (cascade) of such symmetry breaking, bifurcation "episodes" stimulated by the shifting of the numerical values of the control parameters beyond critical thresholds; these broken symmetries increase the complexity of the system by sequentially reducing the number of degrees of freedom of the new pattern(s) (via triggering long-range spatial-temporal correlations). The greater the reduction of symmetry, i.e. the greater the number of constraints a pattern incorporates, the greater the length of the minimum program necessary for the "complete" description of the system - which would lead to its replication.

So we may consider that each pattern (at a hardware level) is associated with a number of digits (at a software level) - we could classify them as beloging to two categories, "pre"-digits and "post"-digits. Pre-digits compose the set of the successive critical (threshold) numerical values of the control parameters, which triggered the corresponding sequence of bifurcations whose final product of broken symmetry is the pattern under consideration. The more abstract the pattern is, the longer the sequence of pre-digits needed to mark (and preserve) its "dotted" or "punctuated" evolutionary history. We may say that the pre-digit set is the evolutionary "zip-code" of the pattern. If the set of pre-digits constitutes the "address" of the pattern, the set of post-digits constitutes its "ID card" or rather its "fingerprints": the set of post-digits contains the set of the amplitudes of the "order parameters" (Haken (1978)) or the dominant eigenfunctions, which emerge after an instability. These eigenfunctions form the collective or macroscopic properties of the pattern concerned.

If a pattern dynamically produced as above is now described or coded using a string of $\{0,1\}$ symbols, the length of the shortest such string can be deduced as follows (see e.g. Schalkwijk (1982)): Let p be the a priori probability of 0's and 1-p that of the 1's in a fair coin tossing realization of length N digits (bits). There are $\binom{N}{p\,N}$ such sequences,

or $$\frac{N!}{[(pN)!]\,[N(1-p)!]}\,.$$

Hence $\log_2\binom{N}{pN}$ bits suffice to uniquely specify such a sequence, or

$\frac{\log_2\binom{N}{pN}}{N}\ \frac{\text{bits}}{\text{digit}}$; using Stirling's formula

$$\xi! \approx \sqrt{2\pi\xi}\ \ \xi^{\xi}\cdot e^{-\xi}\ ,\ \xi \gg 0$$

and applying it above, we get the expression

$$S(p) = -p\log_2 p-(1-p)\log_2(1-p)\ \frac{\text{bits}}{\text{digit}} \tag{2}$$

for uniquely specifying an N sequence of 0 and 1 with p N symbols 0 and $(1-p)N$ symbols 1 (as $N \to \infty$).

However, this is not all: To reconstruct a given <u>finite</u> length N sequence of 0's and 1's where p is the a priori probability of 0's, one needs to specify the number of zeros which, for N finite, is not exactly known; that is, one needs an extra amount of information $\log_2(N+1)$ bits long.[1] So the total <u>complexity</u> of the pattern is

$$N' \sim \log_2(N+1) + \log_2\binom{N}{pN} \simeq \log_2 N+NS(p) \quad \text{for } N \gg 1 \quad \text{or just}$$

$$S(p)\ \frac{\text{bits}}{\text{digit}} \quad \text{as } N\to\infty\ .$$

Chaitin (1975) showed clearly that the overwhelming majority of the above sequence are incompressible, i.e. random.

Indeed, consider the number of sequences of given length N which are com-

---

1) N+1 digits are needed instead of N, just in order to make the difference between N <u>zeros</u> and N <u>ones</u>.

pressible up to K bits, i.e. $N' \leq N-K$. This number equals

$$\Sigma = 2^1 + 2^2 + \ldots + 2^{N-K-1} = 2^{N-K} - 2. \quad (3)$$

The percentage of compressible sequences is thus

$$n = \frac{\Sigma}{2^N} \sim 2^{-K} (N \gg 1). \quad (4)$$

So, for K = 10 for example, only one in a thousand sequences can be specified by an algorithm 10 bits shorter than the sequence itself.

This fact prompts us to suggest a dynamical gadgetry for the memory of our receiver. Instead of storing the full repertoire of the transmitter messages in the memory bank of this receiver (something quite inpractical for large N), one rather implements, at the hardware level of the decoding system, an algorithm, i.e. non-linear recurrent relationship (map), which upon request, i.e. after being triggered from the transmitter of the partner-system, puts forth "here and now" a whole set of aperiodic strings of digits - whose "grammar" is tempered by the shape of the map. Most of them will be pseudorandom sequences and hence the probability of matching any incoming "packet" of given length N will go to one, asymptotically with time.

C. Organization

A cognitive device capable of compressing and simulating a physical phenomenon possesses a more or less advanced degree of "self"-organization.

In general, self-organization is associated with the spontaneous emergence of long-range spatial and/or temporal coherence among the variables of the (organized) system. Nevertheless, the system may be quite "innocent" of this self-organization, namely it may not use it further, or may not know how to use it, or may not know that it knows how to use it... The point is that a self-organized system becomes a cognitive system only beyond a minimum number (namely two) of hierarchical levels; (this means that a one-to-one mapping does not produce cognition).

The sine qua non prerequisite for the emergence of a self-organizing cognitive system is the ability of this self-organizing system to replicate another system (including itself or parts of itself). This ability of self-replication is in turn the indication that the system knows how to compress and simulate. Examples: Some chemical reactions constitute dynamical systems which are self-organizing but are not cognitive. The genome of any biological organism on the other hand is a cognitive system (the simplest one) since via transcription and translation (a process involving two hierarchical levels, the level of nucleicacids and the level of polypeptides) it simulates and eventually reproduces itself. Clearly then, cognition is instrumental in compressing the complexity of another system (including parts of itself) via the formation of collective properties (cross-correlations or statistical cross-moments) - thereby dramatically reducing the initial number of the degrees of freedom of the system under observation. Alternatively, as we will see, an attractor acting on any point of its Basin (set of initial conditions) does de facto the same.

Concluding this introduction we suggest that communication is involved in triggering the emergence of new spatial-temporal patterns via a sequence of instabilities producing bifurcations which lead to broken symmetry, and subsequently in creating and maintaining multiple attractors in dynamical systems, whose attractors categorize and compress such patterns. In particular, in Section 7 we shall try to sketch the dynamical equivalent of an information channel mediating the hierarchical mapping between a set of messages of the "transmitter" and a set of "regimes" of the "receiving" or rather the cognitive apparatus - playing the role of a memory bank. (Let us recall again that an information channel, according to Shannon's "static" information theory, provides a one-to-one mapping between the transmitter and the receiver and is in no way associated with the act of "cognition". The sole role of a classical information channel is the mere copying of the transmitted patterns, as they are, at the receiving end with the greatest possible reliability.)

We shall propose that the set of messages should be identified as a set of initial conditions or the Basin of the attractors in an N-dimensional state space, where N stands for the number of variables mediating the dynamical phenomenon under consideration.

The receiver or, more generally, the cognitive apparatus will be identified as the set of the attractors of much lower dimensionality embedded as invariant compact subsets in the same N-dimensional state space.

Falling into the attractor via (short or long) transients in state space corresponds to the act of progressive signal abstraction (or compression). In what follows we will try to sketch systematically how such activities may be implemented. We will fall short in one aspect however: We will be unable to provide a dynamic[1] model of how the cognitive device reconstructs the initial phenomenon from the compressed versions of environmental time series, that is the time series of the rawdata.

---

1) Algorithmic models of information storage and recall have existed for a long time: In 1968 Longuet-Higgins proposed a device which he called a "holophone", which would record temporal information in exactly the same way that holograms record spatial information. The device records the Fourier transform of the autocorrelation function of the temporal information sequence: When supplied with a "cue" consisting of some part of that

2. Symmetry and symmetry breaking. Information generation

What exactly do we mean by the "laws of Physics"? Physics is an empirical science. People make lengthy observations of a certain phenomenon and then try to "theorise" about it, namely to compress the strings of observation into an essentially simple and brief analytical formula (the "algorithm"). This theorising is eventually implemented by some sort of convolution between the observed time-series and the non-linear dynamics of the (human) brain.

By complementing this algorithm with a variety of extrinsic constraints (initial and boundary conditions) and introducing it as an input into a finite-state machine (e.g. a digital computer) people obtain different scenarios of the same phenomenon - depending on the specific choice of initial and boundary conditions. Therefore the "physical laws" are essentially algorithms leading to simulation outcomes or "models" of large classes of dynamical phenomena. If this is so, then the aim of "reducing" the communication process to the laws of physics leads to an immediate paradox, namely the effort to explain simulation using simulation. At this moment, however, it is wise to push the paradox under the rug and try to proceed. We shall return to it in the last section of the paper, where the concepts and examples of hierarchical processing within the frame of an emergent "context" will perhaps be of some help in elucidating "self-reference".

A physical law is always accompanied by a certain number of symmetries. By degree of symmetry we imply the number of equivalent alternative descriptions. For example, a pattern is symmetrical if, after applying a certain transformation to it, it looks the same as it did before. Or, to put it another way, the greater the degree of symmetry of a system (abstract or concrete) the less the information it contains, the less its complexity, the briefer its description, the more accurate the prediction that we can make about its evolution. For instance, for a closed dynamical system the most probable state is the state of thermodynamic equilibrium, which happens to be the state of fullest possible macroscopic symmetry. No matter which initial conditions we

---

sequence it responds by playing back the remainder of the sequence, starting from the point "cued" - thereby making the temporal information "content addressable".

The proof turns on the fact that after the recording phase, and before the cue, the system preserves a record of the autocorrelation function $A(\tau) = \int R(t)R(t+\tau)\,dt$ of the recorded signal $R(t)$. The playback $P(t)$ in response to the cue $C(t)$ is then found to be proportional to the convolution of $C(t)$ with the autocorrelation function

$$P(t) = \int C(t-\tau)A(\tau)\,d\tau ,$$

which is in turn proportional to $R(t-T_0)$, where T_0 is the (arbitrary) time interval between the cue and the passage cued.

Unfortunately, as Longuet-Higgins and others have shown, the signal/noise ratio of the playback from such holophones is very poor unless the length of the cue is comparable to the length of the entire body of information stored.

start from, no matter what kind of perturbative operation we apply, the above system, with probability one, will asymptotically evolve towards the state of perfect homogeneity. The initial question refers, however, to the symmetry of dynamical laws rather than to patterns (which come into existence after allowing the laws to operate for given time intervals).

Again, a natural law possesses a symmetry with respect to certain sets of operators or transformations if afterwards it appears the same as before.

Now we come to the following question: What is the relation between the symmetries (or the "conservation principles") of a physical law and the symmetries of a pattern generated after allowing this physical law to operate for a certain time interval? The general answer to this fundamental inquiry is still unknown. Broken symmetry of patterns may be due either to intrinsic or to extrinsic causes. Intrinsic reasons have to be attributed to broken symmetry of the laws themselves: (for example the parity violation and time asymmetry, which characterize the weak nuclear force and whose asymmetry is manifested also at the atomic level, have recently been taken see Mason (1984) as an explanation for the broken symmetries in polarity ("up-down") and chirality ("left"-right") of biological macromolecules). Another "sui generis" force is gravity. Among the 4 basic interactions gravity has a special status; other interactions act <u>in</u> space and time - which merely serve as the "setting" for physical events. Gravity, however, is not a force on the passive background of space and time; rather it constitutes a distortion of space-time itself; it is a "curvature" of space-time. Gravity, left unconstrained, leads to "snow-balling" irreversible phenomena associated with the creation of "singularities" accompanied by entropy and information production; it is the only force we cannot screen against.

Aside the above two important cases we argue that in the formation of the degrees of symmetry (or the complexity) of a macroscopic pattern there are extrinsic agents which play significant roles, like the asymmetry of initial and boundary conditions (which are introduced ad hoc to the solutions of our differential equations and determine the numerical values of the constants of integration). By far, however, the predominant reason leading to symmetry breaking in patters is the <u>instabilities</u> which cause <u>bifurcations</u>, i.e. branching-offs in the solutions of our equations.

For example, the polarity and chirality of the biological macromolecules have been attributed, by G.Nicolis and I.Prigogine (1981), to the distortion of the bifurcation diagrams - which, on principle, allow mirror-symmetric outcomes - due to the influence of an extrinsic asymmetry exercised on matter by a gravitational or a (polarized) EM field.

The subject of pattern generation and selection, via bifurcations leading to broken symmetry, is thoroughly reviewed in a forthcoming paper by

G.Nicolis (1985), so we will not deliberate further on this topic.

We will simply offer two brief examples of broken symmetry, which increase the complexity of a pattern or produce information.

The first example is cosmological: How can we account for the progressive differentiation of the universe - which is concomitant with entropy production and information generation? There is an extrinsic asymmetric initial condition, namely the "big-bang":

The metric of the universe dilates with time. If the universe contained only matter (say a perfect monatomic gas) it could be shown (Davies (1977), see also J.S.Nicolis (1985)) that the expansion (or the contraction) of this universe would leave the entropy of the baryonic gas invariant: A material universe starting from equilibrium would stay at equilibrium forever. Similarly, if the universe contained only photons, it could be shown that the expansion (or contraction) would leave the entropy of radiation unaffected, too. A photonic universe starting from equilibrium would stay at equilibrium for ever. However, the coexistence of matter and radiation in the expanding "box" creates the following phenomenon:

Matter cools faster than radiation, and energy tends consequently to flow from the radiation field to the matter in an attempt to equalize temperatures. If the expansion rate is smaller than the temperature equalization, hence the homogenization rate of the mixture, the heat flow will take place under equilibrium conditions and the entropy of the mixture will remain constant.

If, however, the expansion rate is larger than the temperature equalization rate, the heat flow from radiation to matter will take place under non-equilibrium conditions and the entropy will increase: the universe under such circumstances would "lock itself" far from thermodynamic equilibrium. In the early phases of the expansion, the rate of expansion was large but the mixture of matter and radiation were so hot that the gradient of temperature equalization exceeded the gradient of expansion: the early universe was in a state of thermodynamic equilibrium: As time went on the rate of expansion decreased due to the gravitational pulling, but it happened in such a way, as the matter-photon mixture cooled, the rate of temperature equalization dropped faster. So a critical epoch T_c was reached beyond which the mixture locked itself far from equilibrium: Concomitant with the entropy production is the phenomenon of progressive differentiation of the universe from a homogeneous ionized matter-photon (plasma) ball to lumps of matter (galaxies) rushing away from each other (and contracting on an individual basis due to the predominance of gravitational pull in small scales) in a "sea" of long-range propagating radiation. (For further details see J.S.Nicolis (1985)). Thus complexity and information are generated.

Our second example concerns the information production via time symmetry breaking leading to asymptotic stability. We consider the following representative case from Poincaré himself; it is a classical example of a two-dimensional dissipative system capable of being attracted to a new time asymmetrical dynamical regime once expelled from a previously homogeneous state via an instability, in ways which are independent of the initial conditions. Consider the system

$$\frac{dx_1}{dt} = \alpha x_1 + \beta x_2 - x_1(x_1^2 + x_2^2) ,$$

$$\frac{dx_2}{dt} = -\beta x_1 + \alpha x_2 - x_2(x_1^2 + x_2^2) , \tag{5}$$

with a trivial steady state $x_1 = x_2 = 0$.

Let us write $x_1 = r \cos\varphi$ and $x_2 = r \sin\varphi$. Substituting in the original equations, we obtain a system with respect to r , φ , namely,

$$\frac{dr}{dt} = \alpha r - r^3 ,$$

$$\frac{d\varphi}{dt} = -\beta . \tag{6}$$

The second equation gives immediately $\varphi(t) = \varphi_0 - \beta t$.

Let us concentrate on the first equation and determine the steady states; putting $dr/dt = 0$, we have $r_1 = 0$, $r_2 = \sqrt{\alpha}$ $(\alpha > 0)$. These two states are represented in the state space by the origin and a cycle centered in the origin with a radius equal to $\sqrt{\alpha}$ of angular frequency β.

The linearized system is

$$\frac{dr}{dt} = \alpha r ,$$

$$\frac{d\varphi}{dt} = -\beta .$$

It is obvious from the first equation that for $\alpha < 0$, the steady state (0, 0) is stable, and it becomes unstable for $\alpha > 0$ (for $\alpha_c = 0$, the steady state is marginally stable). So let us try to see what happens above the critical value $\alpha_c = 0$, for $\alpha > 0$, i.e. let us try to investigate the stability of the closed trajectory $r = \sqrt{\alpha}$.

Let us slightly perturb this trajectory by putting $r = \sqrt{\alpha} + \delta r$. Substituting in the equation

$$\frac{dr}{dt} = \alpha r - r^3 ,$$

we get

$$\delta \dot{r} = \alpha (\sqrt{\alpha} + \delta r) - (\alpha^{3/2} + \delta r^3 + 3\alpha\delta r + 3\sqrt{\alpha}\delta r^2)$$

$$\simeq \alpha (\sqrt{\alpha} + \delta r) - (\alpha^{3/2} + 3\alpha\delta r) = -2\alpha\delta r \ . \tag{7}$$

We deduce therefore that the trajectory $r = \sqrt{\alpha}$ is asymptotically stable, since the radius r decreases for positive δr and increase for negative δr. We have then a limit cycle, i.e. a circular, asymptotically stable trajectory which attracts all trajectories in its neighbourhood. Plotting the diagram, solution versus control parameter, we see that this attractor involves a broken symmetry, not only because the radius $\sqrt{\alpha} > 0$, but also because the limit cycle runs in time only along one direction. The orbit $r = \sqrt{\alpha}$ is independent of the initial conditions, and its period is intrinsically determined by a system parameter rather than by the initial conditions, as in the case of Hamilton oscillators.

Before closing this section, it is of some interest to investigate the recursive relations giving the "zero-crossing" of the limit cycle along, say, the x_1 axis, as it shrinks to its asymptotic value $r_{\infty} = \sqrt{\alpha}$.

Essentially we are interested in the rate of convergence. From the governing equations in polar coordinates we get, after dividing both members,

$$\frac{dr}{d\varphi} = \frac{(\alpha r - r^3)}{-\beta} \ . \tag{8}$$

The distance between two successive crossings is:

$$\int_{r_{n-1}}^{r_n} \frac{dr}{\alpha r - r^3} = -\frac{1}{\beta} \int_0^{2\pi} d\varphi = \frac{2\pi}{[\beta]} \ , \tag{9}$$

or

$$\frac{1}{\alpha} \log \left. \frac{1}{\sqrt{|\alpha - r^2|}} \right|_{r_{n-1}}^{r_n} = \frac{2\pi}{|\beta|} \ , \tag{10}$$

or

$$r_n^2 = \frac{\alpha}{1 + \exp(-4\pi\alpha/|\beta|)\left[(\alpha/r_{n-1}^2) - 1\right]} \ . \tag{11}$$

The convergence is exponentially fast; indeed, by successively applying the above formula for $n = 1, \ldots$ we get

$$r_n^2 = \frac{\alpha}{1 + \exp(-4\pi\alpha n/|\beta|)\left[(\alpha/r_0^2) - 1\right]} \tag{12}$$

where r_0 is the arbitrary radius from which the process started, i.e. the initial condition. We observe that as $n \rightarrow \infty$, the above formula, as expected, gives $\lim_{n \rightarrow \infty} r_n^2 = \alpha$ independent of r_0, i.e. of the initial conditions.

If, in the above example, we introduce a third degree of freedom, namely a second basic frequency ω_2 (ω_1 being the eigenfrequency of the limit cycle oscillator), we move from the plane which so far has been our state space to a torus in three-dimensional state space.

If $\frac{\omega_1}{\omega_2} = \frac{n}{m}$ where n,m integers, then the trajectory of the system on the surface of the torus will be closed and therefore periodic; its projection onto a cross-section on the torus will be a number of points visited by the orbit on a rotating basis. If, however, the ratio $\frac{\omega_1}{\omega_2}$ is an irrational number (which is far more probable), the trajectory on the surface of the torus will be open and dense; given enough time, the trajectory will pass as near as desired to any point of the surface. The cross section of this orbit will thus be a continuous closed curve.

2.1. Non-periodic attractors

The question arising now is whether in three- or higher-dimensional space there exist attractors more "exotic" than steady states, limit cycles, and 2-dimensional tori. If such attractors do exist, they should first of all possess a volume zero in three-dimensional state space. We can imagine a one-dimensional Poincaré map (i.e. a cross section of the state space manifold with a plane) of such an attractor as an infinite ensemble of non-denumerable, not-interconnected segments of points along a straight line, whose points the attractor visits every time it goes through a "zero-crossing" in a non-periodic "chaotic" way.

Do such structures exist on principle along, say, the normalized interval (0 ÷ 1)? The answer is yes and a well-known example is the Cantor set, which is something between a finite ensemble of points and a continuous curve. Consider the closed segment (0...1), divide it into , say, three intervals, and take away the middle interval by leaving the limiting points 1/3 and 2/3. Repeat this procedure an infinite number of times (e.g. next take away the segment 1/9 ÷ 2/9 excluding the limiting points, and the segment 7/9 ÷ 8/9 likewise). We are left with an irregular, infinite, non-denumerable, and not-connected set of points of measure null. In fact, we can immediately see the last property of the ensemble by calculating the total length of its complement, that is, the part of the inverval (0...1) taken away.

The length of this taken-away part is

$$\frac{1}{3} + \frac{2}{9} + \ldots + \frac{2^n}{3^{n+1}} = \sum_{n=0}^{\infty} \frac{2^n}{3^{n+1}} = 1 \quad . \tag{13}$$

The corresponding (still hypothetical) attractor in three-dimensional space is neither a closed curve nor a surface with well-defined Euclidian dimensionality. It is rather a non-closing, non-intersecting curve imprisoned within a finite (subset of) volume in state space, forming an infinite ensemble of two-dimensional sheets within such a volume. It visits these sheets irregularly and has, along a given direction, on at least one of them, a Poincaré map which constitutes a Cantor or a Cantor-like set.

2.2. Non-integer degrees of freedom: Fractal trajecotries (Mandelbrot (1982))

Suppose now that via a cascade of "broken symmetries" we have succeeded in creating (literally and actually), starting, say, from a straight line or a circle, a pattern with a very irregular envelope curve or a complex trajectory of a dynamical system. We are given then a very irregular (aperiodic) segment of a curve (non-differentiable in an infinite number of points - like, for instance, the trajectory described by the Brownian motion of a particle in a liquid) and we are asked to calculate its length between two specific points A and B, as well as its dimensionality. It is fundamental here to examine what we call the "fractal" properties and the "fractal" dimensionality of a wide category of self-similar[1] curves or processes.

To proceed, we first choose the length of a measuring stick, say G. Then we try essentially to count the numer of equal sides of length G of an open polygon whose corners lie on the curve. If G is small enough, it does not matter whether we start from the end A or the end B of the curve. Thus we get an estimate of the length L(G). The crucial point is that unlike the case of smooth, perfectly differentiable curves with radius of curvature $R_i \geq G$, in our case L(G) depends greatly on G. Consequently, it is necessary to know L(G) for several values of G; it would be preferable to know an analytic formula for the function L(G), say of the form $L(G) \sim AG^{1-D}$, where A is a (positive) constant and D is a constant at least equal to unity. Thus D is called the "fractal" dimensionality of our curve under consideration, and may be a non-integer number.

---

[1] By self-similarity we imply the property of an object whose structure as observed on one length scale is repeated on successively smaller scales.

More generally, in order to understand the nature of fractal dimensionality let us start with the straight line in Euclidian space which has dimension one. Hence, for every positive integer N, the segment $0 \leq x < X$ can exactly be decomposed into N non-overlapping segments of the form

$$\frac{(n-1)X}{N} \leq x < \frac{nX}{N} \quad , \tag{14}$$

where n runs from 1 to N. Each of these parts is deducible from the whole by a similarity of ratio $r(N) = 1/N$.

Similarly, a plane has dimension two. Hence, for every perfect square number N, the rectangular $0 \leq x < X$, $0 \leq y < Y$ can be decomposed exactly into N non-overlapping rectangles of the form

$$\frac{(K-1)X}{N} \leq x < \frac{KX}{\sqrt{N}} \quad ,$$

$$\frac{(\xi-1)Y}{N} \leq y < \frac{Y}{\sqrt{N}} \quad ,$$

where K and ξ run from 1 to $\sqrt{N}$. Each of these parts is deduced from the whole by a similarity of ratio $r(N) = 1/\sqrt{N}$. In general, whenever $N^{1/D}$ is a positive integer, a D-dimensional rectangular parallelepiped can be decomposed into N parallelepipeds deducible from the whole by a similarity ratio $r(N)=1/N^{1/D}$. Thus the dimension D is characterized by the relation

$$D = -\frac{\ln N}{\ln r(N)} = \frac{\ln N}{\ln \frac{1}{r(N)}} \quad . \tag{15}$$

<u>Example:</u> Consider the curve of Fig.2 in the interval (0, 1). It is composed of N = 8 "legs". Replace each of its parts by a curve deduced from the whole drawing through a similarity ratio $r(N) = 1/4$. One is left with a curve made up of N^2 legs of length $(1/4)^2$. Next replace each leg of the new curve by a curve obtained from the whole drawing through a similarity of ratios $(1/4)^2$. The desired self-similar curve is obtained, or rather approached, by an <u>infinite</u> sequence of such steps. Clearly, the total length of such a curve in the unit interval tends to <u>infinity</u>, and its fractal dimensionality is

$$D = \frac{\ln 8}{\ln 4} = 1.5 \quad .$$

Let us now try to calculate the fractal dimensionality of a <u>Cantor set</u>. It will be defined (for any bounded set of points) via the number of small "balls" of size ε needed to cover the set. If this number increases for $\varepsilon \to 0$, like $N(\varepsilon) = \text{const} \ \varepsilon^{-D}$, then D is the "fractal" or Hausdorff dimension of the set.

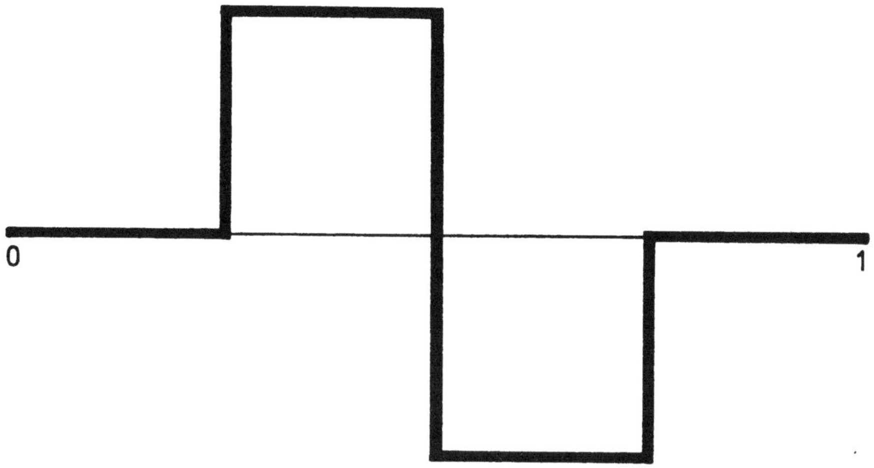

Fig. 2 The formation of a fractal curve through self-similar scaling

We will have again

$$D = \lim_{\varepsilon \to 0} \frac{\ln N(\varepsilon)}{\ln(1/\varepsilon)} \quad . \tag{16}$$

If we are content to know where the set lies within an accuracy ε , then, to specify the location of the set, we need only specify the positions of $N(\varepsilon)$ "spheres" covering the set. Hence the Hausdorff dimension may be interpreted as telling us how much information is necessary to specify the location of the set within a given accuracy. For the Cantor set $\varepsilon = 1/3$ and $N = 2$ for the first division, $\varepsilon = 1/9$ and $N = 4$ for the second division, ... , $\varepsilon = (1/3)^p$, $N = 2^p$ for the p-th division, so

$$D = \frac{\ln 2}{\ln 3} = 0.630...$$

So the fractal dimensionality of the Cantor set is less than its geometrical dimensionality - a necessary (but not sufficient) condition to specify an attractor that is a compressor of information. We note finally that the Cantor set as constructed above has the property of scale invariance, that is, the set between 0...1 looks precisely the same as, for example, the part between 0...1/3, if the latter is examined under a magnifying glass of factor 3.

Before we proceed further in our quest relating essentially shape and stability (thereby heading towards a dynamical theory of information) it is instrumental to provide some crude analogy between hardware and software dynamics.

3. Collisions versus iterations: A basic isomorphism between physical and symbolic dynamics (see also Appendix)

The parallelism between a gas and a recurrent algorithm as far as the mixing property is concerned is a good example of an isomorphism between Hardware and Software, and also provides a connection between physical entropy and Kolmogorov-Sinai entropy (for the latter see Section 6) or between molecular chaos and macroscopic chaos. The fundamental difference, as we will see, is that molecular chaos produces entropy and destroys information, whereas macroscopic chaos produces entropy and information. This is due to the fact that molecular chaos is ergodic whereas macroscopic (dissipative) chaos is ergodic under constraints.

3.1. Molecular chaos

We briefly elaborate on the conditions under which the entropy of a Boltzman gas increases irreversibly with time and examine the premises under which the gas goes to a state of maximum disorder, perfect symmetry or minimum complexity. This state, where all dyadic velocity cross-correlations are δ-behaving, is asymptotically stable.

We write for the entropy the common expression

$$S = - \sum_{i=1}^{\Lambda} P_i \ln P_i \tag{17}$$

where

$$P_i = \lim_{N \to \infty} \frac{N_i}{N} ,$$

N is the total number of particles of the system, Λ is the number of categories or "cells" of particles according to a given partition and N_i is the (average) number of particles belonging to the ith category.

So, for N and N_i large, $i \in (1,\ldots,\Lambda)$, we may write the equivalent expression

$$S = - \sum_{i=1}^{\Lambda} \frac{N_i}{N} \ln \frac{N_i}{N} = - \sum_{i=1}^{\Lambda} \frac{N_i}{N} (\ln N_i - \ln N)$$

$$= \sum_{i=1}^{\Lambda} \frac{N_i}{N} \ln N - \sum_{i=1}^{\Lambda} \frac{N_i \ln N_i}{N} = \frac{1}{N} (N \ln N - \sum_{i=1}^{\Lambda} N_i \ln N_i) . \tag{18}$$

Differentiating with respect to time and considering a closed system so that the total number of "particles" or components of the system N = const, we get

$$\frac{\partial S}{\partial t} = -\frac{1}{N}\sum_{i=1}^{\Lambda}\left(\frac{\partial N_i}{\partial t}\ln N_i + \frac{\partial N_i}{\partial t}\right) = -\frac{1}{N}\sum_{i=1}^{\Lambda}\frac{\partial N_i}{\partial t}\ln N_i \quad , \tag{19}$$

since

$$\sum_{i=1}^{\Lambda}\frac{\partial N_i}{\partial t} = \frac{\partial N}{\partial t} = 0 \quad .$$

Let us now investigate the term $\partial N_i / \partial t$. This rate of change of the number of particles N_i in the compartment i, $i \in (1, \ldots, \Lambda)$, will decrease due to particles leaving i to move to other compartments, and will increase due to the flow of particles into the compartment i from all the others.

Let U_{ij} stand for the probability of having one particle jump from compartment i to compartment j per time unit, and U_{ji} for the probability of a jump from j to i per time unit. Then the number of particles leaving compartment i per time unit will be equal to $\sum_{j=1}^{\Lambda} U_{ij}N_i$, and the number of particles entering compartment i per time unit will be equal to $\sum_{j=1}^{\Lambda} U_{ji}N_j$. So,

$$\frac{\partial N_i}{\partial t} = -\sum_{j=1}^{\Lambda} U_{ij}N_i + \sum_{j=1}^{\Lambda} U_{ji}N_j \quad . \tag{20}$$

Accepting now the initial condition of molecular chaos - namely total lack of correlation[1] between the individual positions and the individual velocities of the moving particles form compartment to compartment - amounts to putting $U_{ij} = U_{ji}$. Hence

$$\frac{\partial N_i}{\partial t} = \sum_{j=1}^{\Lambda} U_{ij}(N_j - N_i) \quad .$$

Substituting this expression for $\partial N_i / \partial t$ in the equation for the rate of entropy change, we get

$$\frac{\partial S}{\partial t} = \frac{1}{N}\sum_{i=1}^{\Lambda}\sum_{j=1}^{\Lambda} U_{ij}(N_i - N_j)\ln N_i \quad , \tag{21a}$$

---

1) This lack of correlation is attributed to cascading collisions among the individual particles - which results in a forgetting of any specific initial conditions (see Appendix).

Interchange of the indices i, j under the accepted chaotic conditions obviously does not change the left member of the above equation, so we can write again

$$\frac{\partial S}{\partial t} = \frac{1}{N} \sum_{i=1}^{\Lambda} \sum_{j=1}^{\Lambda} U_{ji}(N_j - N_i)\ln N_j = -\frac{1}{N} \sum_{i=1}^{\Lambda} \sum_{j=1}^{\Lambda} U_{ij}(N_i - N_j)\ln N_j \qquad (21b)$$

Adding (21a, b) we finally get

$$\frac{\partial S}{\partial t} = \frac{1}{2N} \sum_{i=1}^{\Lambda} \sum_{j=1}^{\Lambda} U_{ij}(N_i - N_j)(\ln N_i - \ln N_j) \geq 0 . \qquad (22)$$

The sign of equality holds in cases where $N_i = N_j$ for all i and j, that is, at the state of absolute homogeneity or again at the state of perfect symmetry - which is the state of "thermodynamic" equilibrium.

3.2. A one-dimensional map in the (unit) interval

Consider next the very simple (piece-wise linear) recursive relation $X_{n+1} = 2X_n$ (mod 1) within the unit inveral. Start iterating from any arbitrary initial value X_0, which, written in dyadic form, is, say, $X_0 = .1110101011011$. Each iteration moves the dot one position to the right. After ξ iterations the value of the number will be

$$X_\xi = 2^\xi X_0 .$$

The initial value X_0 can never be known exactly - even in theory. Why? Because this would amount to asking an infinite number of questions (and getting an infinite number of answers in the yes-no form), that is, it would amount to performing an infinite number of measurements. The entropy of a continuous distribution is infinite. Given that each measurement requires a finite time (commensurate with the switching time or the relaxation time of the particle-composed "decision making" element of the receiver), gaining knowledge of any number X_0 with infinite accuracy would require an infinite amount of time.

Now $(\Delta X_\xi) = 2^\xi (\Delta X_0)$, which means that the initial uncertainty ΔX_0 (however small, yet finite) is amplified exponentially with the number of iterations. If X_0 is known with an accuracy $\sim 10^{-\tau}$, after $\xi_c = \tau \log_2 10$ iterations the initial condition X_0 is completely forgotten[1] and from then on the

---

[1] The complexity of an algorithm has to do with the number of necessary

recurrent algorithm functions as a "pseudo-random generator" producing an essentially incompressible sequence of digits - although the relationship responsible for such a sequence is quite deterministic.

This, in a nutshell, is "chaos": its existence is due to a combination of a finite accuracy of any initial condition and the exponential propagation of this (inevitable) initial error.

Let us next turn to the match discussed in the "chaotic" literature logistic map $X_{n+1} = \alpha X_n(1 - X_n)$, which possess two steady states $X^* = 0$ and $X^* = 1 - 1/\alpha$. The zero value is unstable for $\alpha \geqslant 1$. The value $X^* = 1 - 1/\alpha$ is stable until $\alpha \leqslant 3$. Beyond that value of the control parameter a period doubling bifurcation process starts and terminates at $\alpha = 3.57$ (the so-called accumulation point). At this point the first aperiodic (chaotic) orbit appears and no periodic orbit is stable. This regime is called chaos and in spite of the fact that no orbit is stable, it consists of dense attracting subsets on the interval rather than discrete points (limit cycles). These intervals of traces of the iterating "polygon" on the X-axis form an asymmetric Cantor set.

A few years ago Feigenbaum (1980) discovered, essentially with the aid of a pocket calculator, that the road to chaos via the cascade of period doubling bifurcations is recursive in the following sense.

The succession of the critical values α_c between 3 and ~3.57 is subjected to two "universal" constants: (1) a scaling coefficient $\mu = 2.50290787...$ which gives the "density-packing index" of the critical points X^* between two successive generations (this means that the distance in the unit interval on the X-axis between each bifurcating state and the point 0.5 is μ times smaller than the distance between its "parent" and the point 0.5); and (2) a convergence coefficient $\delta = 4.66920...$ which gives the convergence rate of the critical values α_c between three generations by the following recursive formula

$$\delta = \frac{\alpha_{c_n} - \alpha_{c_{n-1}}}{\alpha_{c_{n+1}} - \alpha_{c_n}} . \qquad (23)$$

When we move from the stable steady state $X^* = 1 - 1/\alpha$ to the limit cycle of period two (beyond $\alpha_c = 3$) the orbits (that is, the first limit cycle of period two) occupy, with probability one, the interval between the two first satellites X_1^* and X_2^* of X^*, whereas the two surrounding portions $0 \div X_1^*, X_2^* \div 1$

---

iterative steps ξ to carry out a given program. This complexity depends on the degrees of freedom N of the system under simulation. For example, for the solution of a system of N linear coupled algebraic equations we have a cubic polynomial dependence $\xi = f(N^3)$ if we apply Gauss' method, or an exponential dependence $\xi = f(N!)$ if we apply the method of determinants. In any case, if $\xi > \xi_c$, a breakdown of simulation takes place (see also Section 8).

are empty (of orbits). However, at each successive pitchfork bifurcation leading to period doubling, the central third of the newly created intervals is not visited by stable orbits (it stays empty) while the two surrounding sub-intervals remain occupied and so on. At the accumulation point $\alpha_c = 3.57$ we have an asymmetric Cantor set with fractal dimensionality ~ 0.538. Feigenbaum's scaling ratio μ imposes a self-similar character on the distribution of points-states on the unit interval. We thus have a set of disjoint intervals with self-similar characteristics in the unit interval and at the accumulation point $\alpha_c = 3.57$. Beyond the first accumulation point there are infinitely many "windows" of parameter values corresponding to an irregular alternation of chaotic and periodic bands. Finally, for $\alpha_c = 4$ the whole unit interval is occupied with orbits and the fractal dimensionality becomes one. For $\alpha = 4$ we can solve the logistic equation exactly by a simple change of variables $X_i = (1 - \cos 2\pi\theta_i)/2$. Then the logistic equation becomes

$$\frac{1}{2}(1 - \cos 2\pi\theta_{i+1}) = 4\left[\frac{1}{2}(1 - \cos 2\pi\theta_i)\ \frac{1}{2}(1 + \cos 2\pi\theta_i)\right]$$

$$= \frac{1}{2}(1 - \cos 4\pi\theta_i)$$

or

$$\theta_{i+1} = 2\theta_i$$

or

$$\theta_i = 2^i\theta_0 \quad . \tag{24}$$

In this case we discover a solution identical with the Bernoulli map. So, in both, physical systems whose particles are subjected to collisions and recurrent algorithms subjected to iterations leading to chaos we are witnessing a mixing phenomenon which is due to the complete forgetting of the initial conditions. For one-dimensional chaotic maps (like the logistic map) around any accumulation point the fractal dimensionality is below one. However, the fact that the fractal dimensionality of such a set of points is below one does not necessarily prove that we have an attractor. For is we have to investigate still another parameter, which provides some measure of the degree of randomness generated by the deterministic equations - a parameter related to orbital stability.

Orbital stability for a given cycle depends on the behaviour of its neighbouring orbits. If points near an orbit converge toward it, then it is stable to small perturbations and is said to be locally stable. An orbit will be attracting, that is, it will be asymptotically stable, if the orbit is locally stable on the average. In the case of one-dimensional maps, these stability criteria are estimated directly from the slope of the map at points visited by an orbit. In particular, if the slope at a point is less than one, nearby points will be brought closer to it at the next iteration of the map. Similarly, if the slope at a point is greater than one, nearby points will be taken farther away under iteration of the map. An asymptotically stable orbit then requires the average of these slopes along the orbits to be less than one. When the average is greater than one, the orbit is unstable; consequently initially small deviations from the orbit will increase under iteration of the map. To put it in a different way, an average slope above one is an indication that the flow generates variety (potential information or entropy), while an average slope below one is an indication that compressibility and dissipation of the generated variety take place; hence relevation of information.

Then sensitivity to small fluctuations, i.e. to some initial uncertainty in the initial conditions (in specifying or measuring a state), is of decisive importance in characterizing the behaviour of a dynamical system. In such cases, the initial uncertainty grows exponentially with time until one can no longer predict the next state of the system. The information about the initial state is lost in a finite amount of time and so, beyond this time, the system is effectively unpredictable. This sensitivity can be considered as a basic feature of chaos.

To measure the average local stability or the rate at which new information is generated by the dynamical flow or, alternatively, the rate at which information about an initial state is lost, the pertinent parameter to use is the "Lyapunov characteristic exponent" λ . The information change per iteration on the map is computed from the slope dF/dx of the map as

$$I = \log_2 \left| \frac{dF}{dx} \right| \text{ bits .} \qquad (25)$$

If $|dF/dx| < 1$, the map at the specific point acts as an information sink, and if $|dF/dx| > 1$, as an information source.

If $P(x)$ is the asymptotic probability distribution of an orbit at a given parameter value α , then the average amount of information change $\langle I \rangle$ is the Lyapunov characteristic exponent for that particular value of the control parameter, namely

$$\langle I \rangle = \lambda(\alpha) = \int_0^1 P(x) \log_2 \left| \frac{dF(x, \alpha)}{dx} \right| dx \, . \qquad (26)$$

If one assumes ergodicity of the orbit within the attractor, there is an alternative form for computing $\lambda(\alpha)$, namely

$$\lambda(\alpha) = \lim_{N \to \infty} \frac{1}{N} \sum_{i=1}^{N} \log_2 \left| \frac{dF(x, \alpha)}{dx} \right|_{x_i} \qquad (27)$$

where N is the number of iterations that the particular orbit undergoes in the map.

Now the calculation of the asymptotic probability distribution of the specific orbit goes as follows: We arbitrarily assume some initial probability density $P_1(x)$ (say equal to one everywhere) along the interval. The resulting value of its first iteration $P_2(x)$ corresponds to the original $P_1(x)$, transformed by the action of the map. The number density of trajectories impinging on some small interval in F(x) is equal to the densities at the inverse points of the mapping weighed by the slope at these points. This is expressed by the law of conservation of probability or the conservation of trajectories. If $x_i = F^{-1}(x)$ is the inverse function of F(x), then

$$P_2(x) = \sum_{i=1}^{2} \frac{P_1(x_i)}{|dF/dx|_{x_i}} .$$

In general, if $x_i^{(n)} = \left[F^{-1}(x)\right]^{(n)}$ is the inverse function of $F^{(n)}(x)$, then

$$P_n(x) = \sum_{i=1}^{2^{n-1}} \frac{P_1(x_i^{(n-1)})}{\left|\frac{dF^{(n-1)}(x)}{dx}\right|_{x_i^{(n-1)}}} \tag{28}$$

Successive iterations of this procedure will produce closer and closer approximations to the correct equilibrium value of P(x). [If the map has a stable periodic solution, P(x) converges to sharp spikes on the periodic orbits, i.e. P(x) will be composed by a series of δ-functions in this case. In the chaotic regime, where only part of the interval is covered, P(x) may also be discontinuous but non-zero over a finite range of x.]

So, with the above definition of the Lyapunov exponent, we see that

<u>for $\lambda(\alpha) < 0$,</u> the orbit is stable and periodic (the iterative process acts as an information sink),

<u>for $\lambda(\alpha) = 0$,</u> the orbit is marginally stable (information is neither produced nor dissipated); and

<u>for $\lambda(\alpha) > 0$,</u> the orbit is locally unstable and chaotic (the iterative process acts as an information source).

In that respect, a <u>steady state</u> in three-dimensional space has three Lyapunov exponents, all being negative (-, -, -); a <u>limit cycle</u> in three-dimensional space has three Lyapunov exponents, two of them negative and one (along the flow) zero (-, -, 0); and a <u>strange attractor</u> is characterized by (+, 0, -), that is, one positive Lyapunov exponent λ_+, one zero λ_0, and one negative λ_-. In other words, a strange attractor creates information along a given direction and compresses information along another direction. Along the flow the Lyapunov exponent is zero.

However, to be an attractor the system must have $\sum_{i=1}^{3} \lambda_i < 0$, i.e. the absolute value of the negative Lyapunov exponent must be greater than the value of the positive Lyapunov exponent. The above sum gives the average volume contraction rate in state space.

4. Lyapunov exponents of chaotic flows in multidimensional state space. The non-uniform factor NUF

The characteristics one deduces about the Lyapunov exponents from one- or higher-dimensional Poincaré maps (difference equations) do not necessarily coincide with the behaviour of the Lyapunov exponents deduced from the "original" dynamical flows, i.e. from the differential equations.

4.1. The concept of the local divergence rate

The spectrum of the Lyapunov exponents constitutes a way to classify dynamical systems in general as well as particular solutions of a dynamical system. This is nowadays widely used in the literature (Haken (1984), Lichtenberg and Liebermann (1983), Shaw (1981)).

In this section we want to interpret the maximal Lyapunov exponent as the time average of a statistical variable, which we call the local divergence rate $y(\underline{x}(t))$ (cf. Nicolis et al. (1983), Grassberger and Procaccia (1983), and Benettin (1983)). Besides the average, further information about the system is contained in the higher moments of the statistical variable like the variance. The variance of the local divergence rate $(\Delta y)^2$ is a measure for the non-uniformity of the dynamical behaviour of an attractor with respect to the separation of nearby trajectories. Besides the Lyapunov exponents, the variance $(\Delta y)^2$ will give additional information about a dynamical system and may turn out to be helpful for a further classification of chaotic systems.

In Section 4.2 we derive the expression for the statistical variable, whose first moment will be the Lyapunov exponent. We will discuss the meaning of the corresponding variance and define the Non-Uniformity Factor (NUF) Δy in the context of dynamics. In Section 4.3 we will consider a representative example analytically.

4.2. Average value and variance of the local divergence rate

We start from a continuous dynamical system, which is described by a set of coupled ordinary non-linear differential equations

$$\dot{\underline{x}} = \underline{F}(\underline{x}) \quad ^{1)} . \tag{29}$$

---

1) Explicit time dependence can be eliminated by increasing the dimension of the system by one.

The maximal Lyapunov exponent is defined as (Haken (1984), Lichtenberg and Liebermann (1983), Shimada and Nagashima (1979), G. Benettin et.al. (1978))

$$\lambda = \lim_{t \to \infty} \frac{1}{t} \ln \| \underline{u}(t) \| \quad ^{2)} , \tag{30}$$

where $\underline{u}(t)$ is a solution of

$$\dot{\underline{u}} = \frac{\delta F}{\delta \underline{x}} (\underline{x}(t)) \underline{u} \quad , \tag{31}$$

which is obtained by linearizing (29) along a trajectory $\underline{x}(t)$ of (29). Note that (31) is a linear system of differential equations with time dependent coefficients that describes the behaviour of a perturbation of the trajectory $\underline{x}(t)$ of system (29). The solution of (31) to a given initial condition $\underline{e}(t=0)$ at time t can be written as

$$\underline{u}(t) = U_0^t(\underline{x}(0))\underline{e}(0) \quad . \tag{32}$$

$U_0^t(\underline{x}(0))$ is a time dependent matrix and is sometimes called fundamental solution matrix. The argument $\underline{x}(0)$ shall indicate that $U_0^t(\underline{x}(0))$ depends on the trajectory $\underline{x}(t)$ with initial condition $\underline{x}(0)$.

We will now write (32) and (20) in a form which is very helpful for the intuitive understanding of the Lyapunov exponent and which is more suitable for the numerical computation of flows as well as for discrete maps (Benettin et.al. (1978)). To this end we formally discretize time into intervals of length τ, such that $t = n\tau$. A solution (32) to (31) can then be written as

$$\underline{u}(t) = U_{(n-1)\tau}^{n\tau} \; U_{(n-2)\tau}^{(n-1)\tau} \; \dots \; U_{\tau}^{2\tau} \; U_0^{\tau} \, \underline{e}(0) \quad .$$

We take τ as the time unit and, changing the notation slightly, we obtain

$$\underline{u}(n) = U_{n-1}^{n} \; U_{n-2}^{n-1} \dots U_1^2 \; U_0^1 \, \underline{e}(0) \quad . \tag{33}$$

Let us read (33) from right to left. $U_0^1 \underline{e}(0) = \underline{u}(1)$ denotes the solution of (31) to the initial condition $\underline{e}(0)$ after time τ. It is a vector with length d_1 and direction $\underline{e}(1)$:

$$U_0^1 \underline{e}(0) = d_1 \underline{e}(1) \quad , \quad \| \underline{e}(1) \| = 1 \quad .$$

---

1) Rigorously, lim should be replaced by lim sup, see Haken (1983).

Now we apply U_1^2 to $\underline{e}(1)$ and obtain another vector of length d_2 and direction $\underline{e}(2)$. Finally we obtain

$$\underline{u}(n) = d_n d_{n-1} \cdots d_2 d_1 \, \underline{e}(n) \quad , \qquad \| \underline{e}(n) \| = 1 \ . \tag{34}$$

We have changed a product of operators into a product of real numbers. The Lyapunov exponent (30) now reads

$$\lambda = \lim_{n \to \infty} \frac{1}{n\tau} \ln \| d_n \cdots d_1 \, \underline{e}(n) \|$$

or

$$\lambda = \lim_{n \to \infty} \frac{1}{n} \sum_{k=1}^{n} \frac{1}{\tau} \ln d_k \quad . \tag{35}$$

From the definition of $d_k = \| U_{k-1}^k \, \underline{e}(k-1) \|$ we see that $\ln d_k$ is the exponential change of the length of $\underline{e}(0)$ during the time interval τ when the system (29) moves along the trajectory between $\underline{x}((k-1)\tau)$ and $\underline{x}(k\tau)$. Normalizing this quantity to one time-unit we define the <u>local divergence rate</u> for discrete or discretized systems

$$y_k = \frac{1}{\tau} \ln d_k \ . \tag{36}$$

Equation (35) now suggests to interpret y_k as a statistical variable. The Lyapunov exponent λ is the time average (in time-units of τ) of the local divergence rate y_k along the trajectory $\underline{x}(t)$:

$$\lambda = \langle y_k \rangle \quad . \tag{37}$$

Note that although (35) describes a discrete sampling of y_k along a continuous flow it is still exact. $y_k = (1/\tau)\ln d_k$ is obtained by solving the continuous equations (29) and (31). On the other hand, (35) can also be used for discrete maps. y_k is then obtained by iterating the map and its linearization.

With the interpretation of λ as the time average of the local divergence rate we define the non-uniformity factor (NUF) to a solution $\underline{x}(t)$ as the corresponding standard deviation

$$\Delta y = \langle (y_k - \langle y_k \rangle)^2 \rangle^{1/2} = (\langle y_k^2 \rangle - \langle y_k \rangle^2)^{1/2} \tag{38}$$

or

$$(\Delta y)^2 = \lim_{n \to \infty} \left(\frac{1}{n} \sum_{k=1}^{n} (\frac{1}{\tau} \ln d_k)^2 \right) - \lambda^2 \quad . \tag{39}$$

The NUF is a measure for the deviation of the local divergence rate from its mean value, the Lyapunov exponent. It characterizes the non-uniformity of a solution of a dynamical system with respect to the sensitivity to initial conditions, or, in other words, how much the local divergence rate changes along the flow. Chaotic flow, e.g., with uniformly turbulent behaviour all over the attractor, yields a positive Lyapunov exponent and a zero variance. Intermittent chaos, however, where regular laminar and chaotic phases alternate irregularly with each other, also gives a positive Lyapunov exponent (possibly the same) but the standard deviation (NUF) should be large, since the local divergence rate varies strongly. So the NUF $\Delta \gamma$ can be a very helpful quantity, besides the Lyapunov exponents, to characterize the behaviour of dynamical systems, especially chaotic ones. Its computation does not take more effort than that of the Lyapunov exponents.

4.3. An example

For a discrete-time dynamical system, which is generated by a map f on the one-dimensional interval I, the Lyapunov exponent λ as defined in (35) is given by

$$\lambda = \lim_{n \to \infty} \frac{1}{n} \sum_{k=0}^{n-1} \ln \| f'(x_k) \| \quad , \tag{40}$$

where $x_k = f^k(x_0)$, and x_0 is some initial value in the basin of the attractor under consideration. In the case of one-dimensional maps we can apply the ergodic hypothesis and replace the time average (40) by the ensemble average of the local divergence rate $\gamma(x)$ with respect to the probability-density-function $p(x)$ of f (Shaw (1981)). In our case the divergence-rate $\gamma(x)$ as defined in (36) takes the simple form

$$\gamma(x) = \ln \| f'(x) \| \quad . \tag{41}$$

The Lyapunov exponent λ therefore can be expressed as

$$\lambda = \langle \gamma(x) \rangle = \int_I p(x)\gamma(x)dx \quad . \tag{42}$$

Accordingly, we obtain the NUF as the standard-deviation $\Delta \gamma$ of the statistical variable $\gamma(x)$(see (38)). The interpretation of the NUF is especially simple in this case: it just describes the variation of the modulus of the slope of the function f along the attractor. Thus, for the "uniform" maps $f(x) = 2x \bmod 1$ and the "tent"-map,

$$f_{0.5}(x) = \begin{cases} 2x, & x \in [0,\ 0.5]\ , \\ 2(1-x), & x \in [0.5,\ 1]\ . \end{cases}$$

We now wish to disturb this uniformity by way of considering piecewise linear but asymmetric maps.

For perturbations of this type we define the family f_α by

$$f_\alpha(x) = \begin{cases} \frac{x}{\alpha}\ , & x \in [0, \alpha]\ , \\ \frac{1-x}{1-\alpha}\ , & x \in [\alpha,\ 1.]\ . \end{cases} \tag{43}$$

For each α we have constant density $p(x) = 1$, and therefore we can calculate both the Lyapunov exponent λ_α and the NUF Δy_α analytically. The Lyapunov exponent is given by

$$\lambda_\alpha = (\alpha - 1)\ln(1-\alpha) - \alpha \ln \alpha\ . \tag{44}$$

From

$$y_\alpha^2(x) = \alpha(\ln \alpha)^2 + (1-\alpha)(\ln(1-\alpha))^2 \tag{45}$$

and

$$\lambda^2 = \alpha^2(\ln \alpha)^2 + (1-\alpha)^2(\ln(1-\alpha))^2 + 2\alpha(1-\alpha)\ln \alpha \ln(1-\alpha) \tag{46}$$

we deduce the NUF

$$\Delta y_\alpha = (\alpha(1-\alpha))^{1/2} \ln(\frac{1-\alpha}{\alpha})\ . \tag{47}$$

For further details concerning the NUF of other maps and flows see Nicolis et al. (1983).

5. A typical three-dimensional strange attractor. The Lorenz model

The combination of sensitivity to initial conditions with only approximate knowledge about these initial conditions leads to an inability to make accurate long-term predictions about the evolution of systems composed by even three non-linearly coupled variables. From the practical point of view, one of the most important sensitive systems is the atmosphere. Lorenz suggested in 1963 that its dynamics are severely sensitive to initial conditions. The implications of this for weather prediction are quite serious - even if massive improvements in modelling and data collection were made.

And Lorenz still found that by severely truncating the Navier-Stokes equations (describing a system with an infinite number of degrees of freedom) one could end up with a non-linear system with just three variables preserving a lot of the characteristics of the initial system. This "compressed" model we intend to study now in some detail because it happens to be very representative of three-dimensional systems possessing strange attractors and being capable of an unbelievably rich behavioural repertoire.

Let us consider the system

$$\frac{dX}{dt} = \sigma Y - \sigma X \quad ,$$

$$\frac{dY}{dt} = -Y + rX - XZ \quad , \tag{48}$$

$$\frac{dZ}{dt} = -bZ + XY \quad ,$$

where σ, r, and b are control parameters; what X, Y, Z stand for is of no importance in the present context.

The behaviour of the system was first investigated numerically by Lorenz in 1963, for $r = 28$, $\sigma = 10$, and $b = 8/3$. Lorenz found that the system starts a rotation around one of the (unstable) focuses with amplitude increasing with time, thereby forming a divergent spiral. After a number of such oscillations, the system suddenly leaves this regime and goes monotonically towards the second available focus, around which it starts again an oscillatory motion along a divergent spiral. Again, after a certain number of oscillations around this focus, the system jumps anew towards the vicinity of the previous focus, from which it starts again a divergent oscillatory trajectory, and so on. The interesting thing is that the time intervals the system spends in the vicinity of each focus before jumping into the vicinity of the other focus are stochastically distributed, and there is no regularity, whatsoever, in the process, which nevertheless is created by the unfolding of a deterministic (non-linear) dynamics. Also, the number of divergent oscillations the system undergoes around each of the two available foci is apparently irregular and therefore totally unpredictable.

Takeyama (1978) has shown that the Lorenz model of convective instabilities can be reduced to a one-dimensional dynamics of an overdamped nonperiodic oscillator with memory in a potential well. When the potential has double minima, memory controlled oscillations can lead to apparently erratic behaviour. This provides a mechanism through which the accumulated dynamics of the past (a collective property of the single variable) excites and controls the present. Eliminating Y and solving Z in terms of X^2 in the above Lorenz system we get, by Takeyama, the following equation for X(t)

$$\ddot{X} + (1 + \sigma)\dot{X} + \sigma\left\{1 - r + \frac{1}{2\sigma}X^2 + \left(1 - \frac{b}{2\sigma}\right)\int_0^\infty \left[X(t - \tau)\right]^2 \exp(-b\,\tau)\,d\tau\right\}X = 0 \tag{49}$$

We assume that the motion started at $-\infty$ and for $b > 0$ we split $\exp(-b\,\tau)$ into a δ-function and the deviation from it and get

$$\ddot{X} + (1 + \sigma)\dot{X} + \frac{dU}{dX} + \left\{\left(\sigma - \frac{b}{2}\right)\int_0^\infty \left[X^2(t - \tau) - X^2(t)\right] e^{-b\tau}d\tau\right\} X = 0 \tag{50}$$

where

$$U(X) = \sigma\left(\frac{1 - r}{2}X^2 + \frac{1}{4b}X^4\right) .$$

This equation describes the motion of an overdamped particle into a potential well U(X) under the action of an additional force which depends on the history of motion. This force sets the context of motion.

For $r < 1$ the potential well has one stable minimum $X = 0$. At $r = r_c = 1$ we witness a bifurcation and for $r > 1$ the well possesses two stable minima $X = \pm\sqrt{b(r - 1)}$ (see Fig. 3).

The cumulative memory term has a destabilizing effect on the two steady states if $\sigma > b/2$, and a stabilizing effect if $\sigma < b/2$. In other words, for $\sigma > b/2$ the history-dependent force acts so as to amplify any original perturbation. Instability (and jumping onto the other well) happens then when this context- or memory-depending force, which "feeds energy" into the system, overcomes the inhibition due to the viscous term $(1 + \sigma)\dot{X}$. The result is an irregular sequence of jumping to the right (0) or to the left (1) well, that is, the production of a Markov chain or an essentially pseudorandom sequence of 0's and 1's.

The past history of the motion does not necessarily represent itself via the amplitude of the last oscillation but rather by some averaging property of many oscillations of the past. In that sense the Markov chain generated may have a long term memory in general.

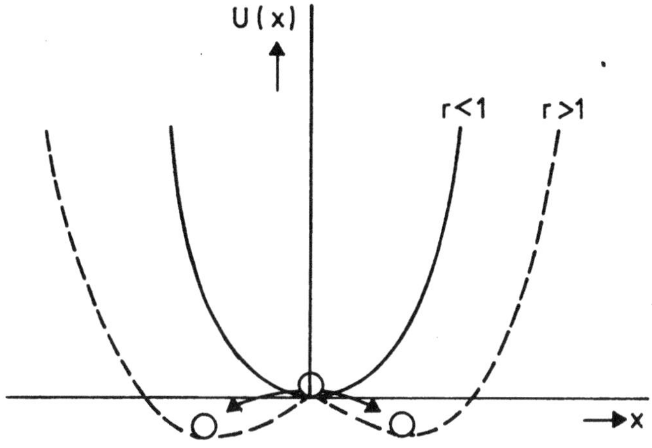

Fig. 3 The potential U(X) for $r < 1$ (continuous line) and for $r > 1$ (dotted line)

Fig. 4 gives the projections of the trajectories of the Lorenz system for the above values of the parameters in state space, on the X-Y and X-Z planes, respectively. In a certain sense it reminds us of the "flow" of the communication transaction sketched in Fig. 1.

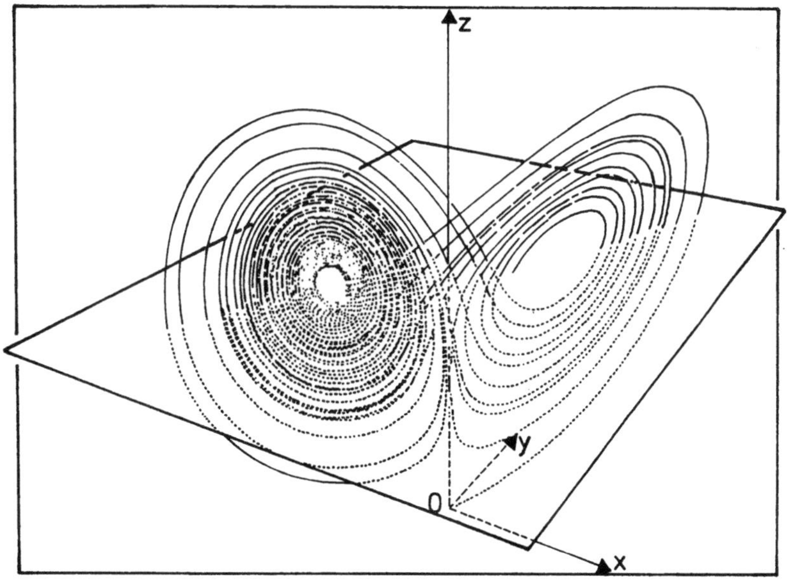

Fig. 4 The trajectory of the Lorenz attractor on the X-Y plane, and on the X-Z plane ($r = 28$, $\sigma = 10$, $b = 8/3$)

The "chaotic" behaviour shown in this figure is just one out of many in a rich repertoire of different modes of chaos that the simple Lorenz model is capable of. Indeed, for higher values of the control parameter r, a succession of new dynamical regimes appears - always investigated via computing simulation. We limit - and end - the discussion on the Lorenz system properly by giving a brief description of two most interesting phenomena, namely "intermittency" and "metastable" chaos. At $r \sim 24.74$ the system displays a "metastable" chaos and a hysteretic behaviour.

To see this more clearly, consider a sketchy one-dimensional Poincaré map performed on the Z-Y plane of the attractor (Fig. 5). A portion of this map has a slope less than one, and it has a stable fixed point at the origin. Any trajectory falling at points less than x_0 will be attracted to the origin. However, a trajectory initiated above x_0 may move chaotically for quite a while before becoming trapped in the region between zero and x_0. The parameter r can be adjusted to make the trapping region $0 \div x_0$ as small as desired, resulting in arbitrarily long "half-lives".

Metastable chaos can then be described as a strange attractor with a finite probability of decaying into a fixed point or to a limit cycle; it implies a

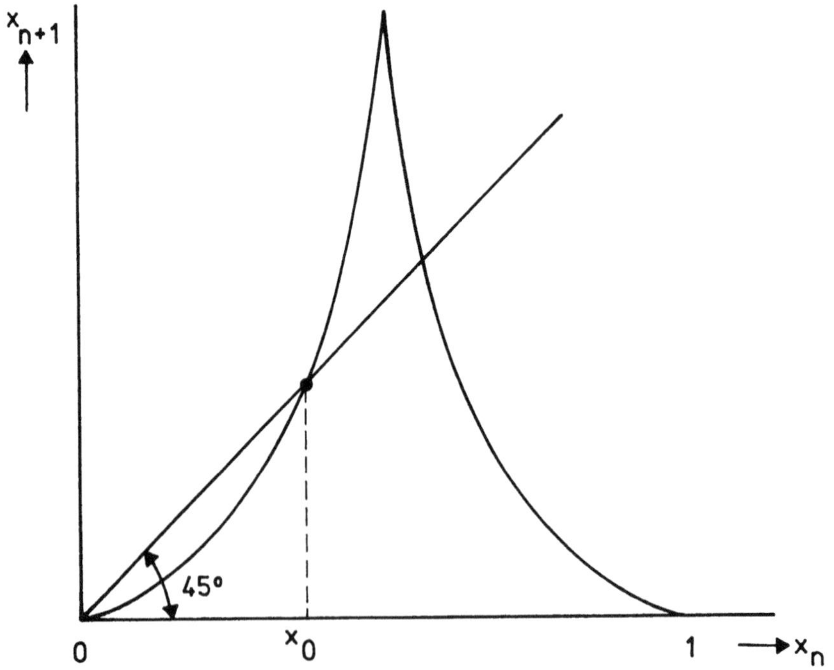

Fig. 5 Sketch for the one-dimensional (cusp-like) map of the Lorenz attractor, illustrating the conditions for the existence of metastable chaos

transitory regime whose decay times have an exponential distribution. Almost every trajectory eventually settles down to a periodic orbit after some initial period of irregular oscillation. The mean duration of the chaotic regime is not long in general, typically about 50 iterations, but by trial and error on the map initial conditions which remain chaotic much longer can be found. In other situations a metastable chaotic regime might decay to a steady state or to one which is also chaotic, but which has significantly different behaviour, i.e. has a different probability density function and occupies a different sub-interval of the whole attractor. On the other hand, the new regime might also be metastable, eventually decaying to a third regime or possibly back to the original one.

The strange aperiodic attractor properly appears for $r > 24.74$ and then slowly evolves up to $r \sim 145$. From this point on, a complicated sequence of changes occurs in the topology of the attractor until its ending at $r = 148.4$, where a stable limit cycle sets in (Manneville and Pomeau (1980)).

Beyond $r_c = 166.07$, the limit cycle to which the strange attractor degraded becomes unstable, and the system displays a new dynamical behaviour called "intermittency". Loosely speaking, intermittency (realized for some windows of the control parameter r in the Lorenz system) refers essentially to the physical inability of achieving perfect entrainment or stable, strict locking between the phase of a dissipative non-linear oscillator and the phase of an external periodic excitation. By the time the limit cycle in state space is about to close, a "scrambling" process, as it were, intervenes and destroys the orderly trajectory. For a brief time interval it creates a chaotic motion amounting to broadband noise or to a broadening of peaks in the power spectrum, and then the building up of a limit cycle appearance, and so on, the time sequence of intermittent points being essentially random.

More formally, we can understand the principle of intermittency by resorting to the Poincaré map of the Lorenz attractor around the value $r \sim 167$ of the control parameter, in Z-Y plane as before.

As sketched in the detailed Fig. 6a, as long as the curve of the map intersects the first bisectrix, we have a <u>stable</u> periodic motion, but as we move up the curve, the <u>two</u> fixed points (one stable, one unstable) first collapse and then disappear.

Just after this disappearance (Fig. 6b), and as long as the curve stays nearly tangent to the first bisectrix, the system keeps memory of the formerly stable fixed point, and we observe an almost stable limit cycle except at the very beginning and at the very end. Then, far enough from the old fixed point, as the iteration proceeds starting from the left, the representing point enters a kind of "channel", slows down, goes through the channel, and then speeds up and leaves the channel on the right. Then, far enough from the old fixed point, the system enters a "turbulent" regime, during which correlations concer-

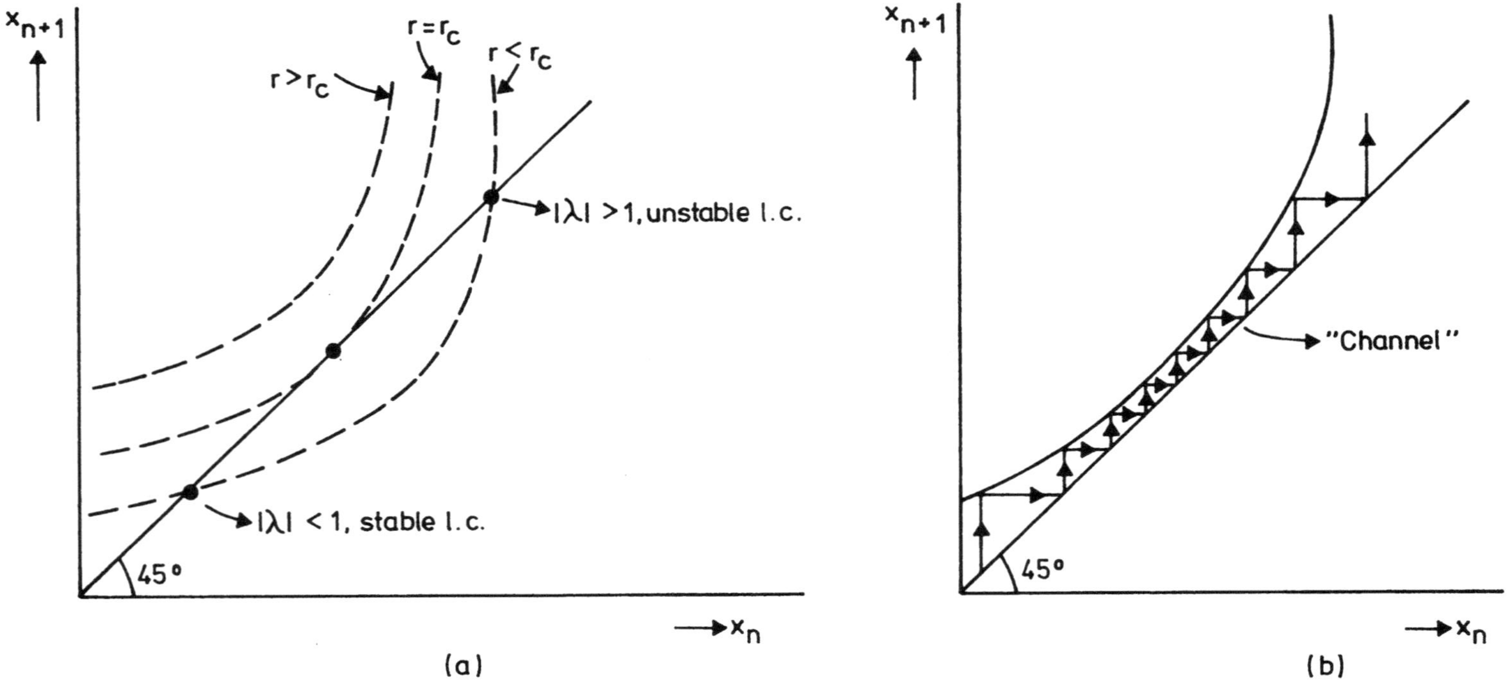

Fig. 6a, b Detailed conditions for the onset of intermittency regime ning phase locking are broken and the system randomly explores a large portion of the state space. This bursting phase ends up sooner or later when the moving point, as it explores the state space, comes close to the formerly stable limit cycle and so "reinjects" the trajectory in the vicinity of the old fixed point.

Due to a certain randomness in the reinjection process, the channel is not always covered in its entire length in the vicinity of the old fixed point, and there may be a certain dispersion in the duration of the "laminar" intervals. In conclusion, then, the intermittent behaviour is merely a consequence of the translation of the parabolic part of the Poincaré map (as the control parameter increases) so that it first intersects, then runs tangent to, and finally loses contact with the bisector line.

5.1. The rate of information production by the Lorenz attractor ($\sigma = 10$, $b = 8/3$, $r = 28$)

In order to calculate the information generated by this dynamical system we first construct a Poincaré map on the Z-X projection of the attractor by making a transversal cut to the flow and then recording successive passes through a maximum along, say, the z-direction. As the attractor is symmetrical with respect to the transformation $x \to -x$, $y \to -y$, $z \to z$, the cut on one "leaf" will yield the same map as a symmetrically placed cut on the other.

The time $t(x)$ between successive passes through the Poincaré map is also recorded (Shaw (1981)). Having the return map $F(x)$ and the function $t(x)$, we next calculate the probability density $P(x)$ using the iteration method developed during the treatment of the logistic map. The information production is then equal to

$$\langle I \rangle = \int_0^1 P(x) \log \left| \frac{dF(x)}{dx} \right| dx \sim 0.98 \text{ bits per iteration .} \qquad (51)$$

This result may be compared with the corresponding value of $\langle I \rangle$ for the logistic map and $\alpha = 4$.

It can be deduced [1)] that for this extreme value of the control parameter,

$$P(x) = \frac{1}{\pi \sqrt{x(1-x)}} , \qquad (52)$$

---

1) This is done as follows: Introducing the transformation $x' = (2/\pi) \sin^{-1} \sqrt{x}$ on the logistic map $F(x) = 4x(1 - x)$, we come up with the symmetrical "tent" map:

and so

$$\langle I \rangle = \frac{1}{\pi} \int_0^1 \frac{\log_2 \left[4(1 - 2x)\right]}{\sqrt{x(1 - x)}} dx = 1 \text{ bits per iteration} \tag{53}$$

- something we should expect since for $r = 4$ the logistic map is exactly two onto one. For the values $r = 28$, $b = 8/3$, $\sigma = 10$ of the control parameters the cusp-shaped Lorenz Pointcaré map is nearly symmetrical and almost two onto one, but not quite.

In order now to describe the information-production characteristics of the Lorenz attractor, we calculate the expression

$$\left\langle \frac{dI}{dt} \right\rangle = \int_0^1 \frac{P(x)}{t(x)} \log_2 \left| \frac{dF(x)}{dt} \right| dx \quad \text{(bits/sec)} , \tag{54}$$

or, in case $P(x)$ is not known, the expression

$$\left\langle \frac{dI}{dt} \right\rangle = \lim_{n \to \infty} \frac{1}{n} \sum_{i=1}^{n} \frac{\log_2 \left| \frac{dF(x)}{dx} \right|}{t_n} , \tag{55}$$

where t_n is the length of time taken by the nth pass. The information production rate for the Lorenz attractor was calculated (Shaw (1981)) and was found to be ~ 1.19 bits/sec.

It is time now to direct our attention more systematically to the study of the parameters which may qualify a strange attractor as an information processor. Obviously, we expect that for values of the control parameters below the initiation of chaos (for which only steady states and periodic orbits exist), the trajectories in state space will only attract, and the corresponding iteration, in the Poincaré map on the interval, will monotonically converge, giving probability densities which are just a set of δ-like functions. Under such subcritical cir-

---

$$F'(x') = \begin{cases} 2x' , & 0 < x' < \frac{1}{2} , \\ 2(1 - x') , & \frac{1}{2} < x' < 1 . \end{cases}$$

Obviously, for this map $P'(x') = 1$. From the conservation of probability $P'(x')dx' = P(x)dx$, we next get

$$P(x) = \frac{dx'}{dx} = \frac{1}{\pi} \frac{1}{\sqrt{x(1 - x)}}$$

which concludes the proof.

cumstances $\langle I \rangle < 0$, i.e. the system will behave asymptotically as an information sink, that is, as an agent which constraints variety, thereby revealing information - essentially, then, a source of "negentropy".

Beyond the critical value(s) of the control parameter(s) for which continuous chaos is established (almost all orbits are unstable), $\langle I \rangle > 0$ along some direction(s) in state space, thereby creating variety or increasing entropy (source(s)). Also, $\langle I \rangle < 0$ along some other directions in state space, thereby dissipating variety or revealing information (sink(s)). It is out of this information-production-dissipation interplay that one expects to derive the salient features which characterize the dynamical system as a more or less efficient information processor.

6. Parameters characterizing the average behaviour of chaotic strange attractors (dimensions, entropies, and Lyapunov exponents)

Three basic concepts are of crucial importance (and need therefore a quantitative measure) in characterizing the dynamical properties of strange attractors:

a) The degree of "compressibility" displayed by the system (which acts as an efficient "vacuum cleaner", so to speak, in functional space).

b) The exact way the entropy is generated through the passage (and amplification) of the intrinsic fluctuations of the system from microscopic to macroscopic hierarchical levels, as the resolution produced by cascading iterations increases.

c) The stability properties of the attractor or the rates of variety generation along given directions, and information dissipation along other directions in state space.

In this section we intend to search for pertinent parameters quantifying for the above average characteristics of chaotic attractors.

We are involved with dynamical systems in discrete time n such as "maps" $x_{n+1} = F(x_n)$, or continuous flows, i.e. ordinary first-order non-linear differential equations $d\underline{x}(t)/dt = \underline{f}(\underline{x}(t);\ \mu)$, where in both cases $\underline{x}$ is generally a vector in a multidimensional state space. Given an initial condition (a value of $\underline{x}$ at $n = 0$ for the map, or $t = 0$ for the continuous system), an orbit is generated $\underline{x}_1, \underline{x}_2, \ldots, \underline{x}_n$ for the map, or a trajectory $\underline{x}(t)$ for the differential system.

Here we are interested in attractors manifested in such systems. An attractor is a compact subset $\underline{C}$ of the state space, with the property that it attracts initial conditions from a region around it once that transients have subsided. "Attraction" here means that for every initial condition, the limit set of the orbit, as n or t go to ∞, is the compact set $\underline{C}$. Every trajectory in this neighbourhood of $\underline{C}$ passes, then, arbitrarily close to every point of the set $\underline{C}$. An attractor is "chaotic" if nearby orbits diverge exponentially from one another with time. In such cases we speak of "sensitive dependence to initial conditions". An attractor, on the other hand, is "strange" if it possesses a non-integer fractal dimension. So the characterization "chaotic" refers to the dynamics of the attractor, whereas "strangeness" refers to the topology of the attractor. The condition for attraction is

$$\nabla\cdot\underline{f} < 0 \qquad \text{or} \qquad \sum_{i=1}^{N} \frac{d\dot{x}_i}{dx_i} < 0 \ . \tag{56}$$

The dimension of an attractor is the first necessary parameter to characterize its compressibility properties. The dimension gives the amount of information necessary to specify the position of a point on the attractor within a given accuracy. Alternatively the information dimensionality can be considered as the number of bits one can dynamically store on the attractor. For trajectories on the attractor, the reduction of dimensionality brings about an accompanying reduction in the information needed to specify an initial condition.

Knowledge of the information dimension of an attractor allows an observer to estimate the information gained when a single, isolated measurement is made at a given level of precision.

When dealing with a dissipative dynamical system, we may begin with an Euclidian state space of initial conditions of large or even infinite dimensionality. After some time, however, the transients die out, and the point in state space that describes the state of the system approaches the attractor. The number of degrees of freedom is reduced in the sense that the number of independent variables inherent to the motion is much less than the number of independent variables needed to specify an arbitrary initial condition. The difference between the Euclidian dimensionality of the state space, where the dynamical system is embedded, and the information dimensionality of the attractor gives, loosely speaking, the average degree of "compressibility" manifested by the attraction.

For simple attractors, defining and investigating the information dimension is trivial. For example, a stationary, time-independent equilibrium (steady state) has dimension one, and a 2-torus has dimension two. In these simple cases the dimension takes values that are integers. This, however, is not the case for strange attractors.

In order to appreciate the properties of a chaotic attractor, we must take into account not only the set itself but also the distribution or density of the points on the attractor. (The reason is that at any fixed level of precision, most information about initial conditions is lost in a finite time T, which will be calculated in Section 9. For times greater than T, knowledge of the future is limited to the information contained in the probability distribution of points on the attractor.)

6.1. The concept of information dimension

The notion of information dimension is intimately liked to the measuring process. Consider a measuring instrument with a uniform scale of resolution ε, so that the measurement of any of the N variables yields one of $1/\varepsilon$ possible numbers. If such a measuring instrument is assigned to each of the N variables of our dynamical system, the quantization of the scales of those instruments induces a partition in the state space, that is, a set of non-empty non-intersecting measurable "hypercubes". The region of the state space containing the attractor is therefore divided into ε^{-N} boxes, each of dimension N. The collection of these boxes is the partition, and each measurable state corresponds to a box. Let $n(\varepsilon)$ be the number of boxes covered by the attractor with non-zero probability (that is, containing at least one point of the attractor). Some boxes will contain more points than others since certain regions in state space are visited by the flow far more often than others. The "natural measure" of a given region is proportional to the frequency with which it is visited. Following Farmer (1982) let $P_i(\varepsilon)$ be the "density probability" of the occurrence of the attractor somewhere on the ith box. The collection $P_i(\varepsilon)$ is referred to as a "coarse-grained" probability distribution of resolution ε. The (average) information contained in a single isolated measurement made at resolution ε is then

$$I(\varepsilon) = - \sum_{i=1}^{n(\varepsilon)} P_i(\varepsilon) \log_2 P_i(\varepsilon) \text{ bits.} \tag{57}$$

As the scale of resolution ε decreases, the number of boxes increases and we get a sequence of successively more refined, less coarsely grained probability distributions. The information dimension D_I is the rate at which the information scales, as the precision of measurement, are increased, namely

$$D_I = \lim_{\varepsilon \to 0} \left[\frac{I(\varepsilon)}{|\log_2 \varepsilon|} \right] . \tag{58}$$

D_I is then given by the asymptotic value of the slope of the graph $I(\varepsilon)$ versus $\log_2(1/\varepsilon)$.

So, if the information dimension of an attractor is known, the information $I(\varepsilon)$ contained in a measurement of a state of resolution ε is estimated as

$$I(\varepsilon) = D_I \left| \log \varepsilon \right| . \tag{59}$$

If the probability of all the boxes in the partition is equal, then

$$I(\varepsilon) = \log_2 n(\varepsilon) , \tag{60}$$

and the information dimension reaches its maximum value, which coincides with the fractal dimension

$$D_F = \lim_{\varepsilon \to 0} \frac{\log_2 n(\varepsilon)}{\log_2 \frac{1}{\varepsilon}} . \tag{61}$$

The fractal dimension is then an upper bound of the information dimension.

What limits ε in practice is, of course, the level of intrinsic (internal) fluctuations in the system and the measuring instrument - collectively labelled as "observational noise". This noise is manifested only during the measuring process and does not affect the phenomenological dynamics of the system. In the above definition of D_I or D_F, external noise is excluded.

6.2. The concept of characteristic Lyapunov exponents and their relation to information dimension

We have already appreciated that the main difficulty in calculating the information dimension comes from the probability density function and particularly the way it scales as the resolution increases to infinity. For the logistic map, for example, this probability density is calculated from the recursive relation

$$P_n(x) = \sum_{i=1}^{2^{n-1}} \frac{P_1(x_i)}{\left|\frac{dF^{(n-1)}}{dx}\right|_{x_i}} . \tag{62}$$

where $x_i = \left[F^{(n-1)}\right]^{-1}$.

Therefore, in essence, the difficulty springs from the fact that the Jacobian matrix or determinant of the map (or the attractor in general) is not constant, that is, it changes with the order of the iteration. In fact, this is also the main reason causing the difference between the information dimension and the fractal dimension.

In turn, the Jacobian matrix is related to the stability properties of the attractor. So we are tempted to conjecture that the dimensions of chaotic attractors can be determined directly from the dynamics of the system, or in terms of the "spectrum of Lyapunov exponents", of the attractor. (We have already given two examples in the previous section.)

The spectrum of Lyapunov characteristic exponents provides a qualitative estimation of the local stability properties of an attractor. The local stability properties of a system are determined by the response the system manifests under small perturbations. A dynamical system can be stable to perturbations along certain directions and unstable to perturbations in others. All possible perturbations can be examined simultaneously by following the evolution of an ensemble of points that is initially contained in a small N-dimensional sphere, where N is the state-space dimension in which the attractor forms a subset.

Imagine, then, an initially infinitesimal ball that has a radius $\delta(0)$ at time $t = 0$. As the ball evolves under the action of a non-uniform flow, it eventually will distort. Since we have assumed an infinitesimal state-space element, the change in shape, we presume, will be determined only by the linear part of the flow - and it remains an ellipsoid as it evolves.

Let $\delta_i(t)$ be the ith member of the set of the principal axes of the ellipsoid at time t . The spectrum of Lyapunov exponents λ_i for a given starting position* is determined as

$$\lambda_i = \lim_{t\to\infty} \lim_{\delta(0)\to 0} \frac{1}{t} \log_2 \left[\frac{\delta_i(t)}{\delta(0)}\right] . \tag{63}$$

There are N Lyapunov exponents in the spectrum of an attractor of an N-dimensional dynamical system. Positive Lyapunov exponents measure the average exponential spreading of nearby trajectories and negative exponents measure the average exponential convergence of trajectories onto the attractor. The sum of the Lyapunov exponents is the average divergence, which for a dissipative system (possessing an attractor) must always be negative. (For a Hamiltonian system this divergence is zero.) For a number of dissipative systems, there are examples where computing simulation indicates that the values of Lyapunov exponents are invariant for all tried initial conditions. If this is so, then the spectrum of Lyapunov exponents may be taken to be a property of the attractor.

It is customary to arrange the Lyapunov exponents in decreasing order. So, for instance, the symbolism +, 0, -, indicates an attractor in a three-dimensional state space with (on the average) exponential expansion along, say, the $\hat{x}$-direction, neutral stability along the flow (say the $\hat{y}$-direction) and exponential contraction of trajectories along the $\hat{z}$-direction. It is important to realize that attractors, unlike stable steady states always have at least one exponent equal to zero since, on the average, points along a trajectory confined to a compact set can neither separate nor merge. Also, for a dissipative system, $|\lambda(-)| > \lambda(+)$.

Consider now a map $\underline{x}_{n+1} = F(\underline{x}_n)$ of the attractor, where $\underline{x}$ is a $\underline{p}$-dimensional vector. The definition elaborated above suggests that the Lyapunov exponents of the discrete system are related to the eigenvalues of the Jacobian matrix of the map (in the case of a stable steady state, the Lyapunov exponents are the real parts of eigenvalues of the interaction matrix).

Specifically, let

$$J_n = J(x_n) \cdot J(x_{n-1}) \cdot \ldots \cdot J(x_1) , \qquad (64)$$

where

$$J(x_1) = \left(\frac{\partial F}{\partial x} \right)_{x_1} ,$$

$$J(x_{n-1}) - \left(\frac{\partial F}{\partial x} \right)_{x_{n-1}} ,$$

$$J(x_n) = \left(\frac{\partial F}{\partial x} \right)_{x_n} .$$

Then

$$J_n = \left(\frac{\partial F}{\partial x} \right)_{x_1} \cdot \left(\frac{\partial F}{\partial x} \right)_{x_2} \cdot \ldots \cdot \left(\frac{\partial F}{\partial x} \right)_{x_n} = \left(\frac{\partial F^{(n)}}{\partial x} \right)_{x=x_0} , \qquad (65)$$

where $x_i = F^{(i)}(x_0)$. Here we make use of the known fact that the slope of the nth iterated map at some point x_0 equals the products of the slopes of the original map at the points $x_1, x_2, \ldots, x_n$.

Let $\sigma_1(n) \geq \sigma_2(n) \geq \ldots \geq \sigma_p(n)$ be the magnitudes of the eigenvalues of J_n. The Lyapunov numbers are

$$\lambda_i = \lim_{n\to\infty} \left[\sigma_i(n)\right]^{1/n} \quad , \quad i = 1,2,\ldots,p \quad , \tag{66}$$

where the positive real n^{th} root is taken. The log of λ_i above gives the "Lyapunov exponents". For the one-dimensional case, the above definition coincides with the result we found in the Section 3.2 of the logistic map if we take the algorithm above, namely

$$\lambda = \lim_{n\to\infty} \frac{1}{n} \sum_{i=1}^{n} \left|\frac{\partial F}{\partial x}\right|_{x_i} \tag{67}$$

for the Lyapunov exponent.

Let us give an example in the case of a <u>two-dimensional</u> map and try to derive a heuristic relationship between the spectrum of Lyapunov exponents and the information dimension from this example. For a two dimensional map, let λ_1, λ_2 be the average distorting coefficients of an infinitesimal circular disk of radius ε (Fig. 7).

Since the system is discrete, the Lyapunov exponent for which divergence takes place corresponds to, say, $\lambda_1 > 1$, and the Lyapunov exponent for which shrinking takes place corresponds to $\lambda_2 < 1$. After n iterations of the two-dimensional map, the initially small circle of radius ε will be transformed into an ellipse with major and minor halfaxes $\varepsilon(\lambda_1)^n$ and $\varepsilon(\lambda_2)^n$, respectively. Now let us try, following Farmer et al. (1982), to relate this spectrum of Lyapunov exponents to the information dimension of the map. We start, as usual, by covering the two-dimensional attractor with $n(\varepsilon)$ squares of

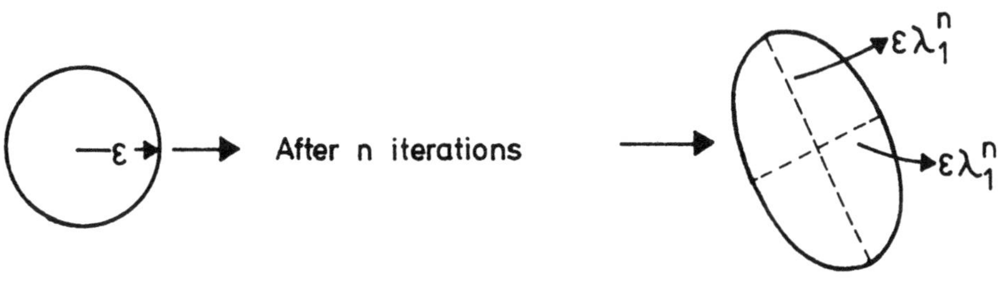

Fig.7 Understanding the action of Lyapunov exponents on a two-dimensional system

side ε . Now let us iterate the map ξ times. For ε small enough (high resolution), the action of the mapping is practically linear over each square. So each element of the partition will be stretched into a parallelogram. The average dimensions of these parallelograms will be $\lambda_1^{\xi}\varepsilon$ and $\lambda_2^{\xi}\varepsilon$, respectively ($\lambda_1 > \lambda_2$). Suppose now we increase the resolution (we make the partition finer) by using squares of side $\lambda_2^{\xi}\varepsilon$. To cover each papallelogram, we used ($\lambda_1/\lambda_2)^{\xi}$ smaller squares, that is, the number of the elements of the new partition will be

$$n(\lambda_2^{\xi}\varepsilon) = (\frac{\lambda_1}{\lambda_2})^{\xi} \cdot n(\varepsilon) . \tag{68}$$

Let us put now $n(\varepsilon) \cong (1/\varepsilon)^d$ and substitute this above. We get

$$(\frac{1}{\lambda_2^{\xi}\varepsilon})^d = (\frac{\lambda_1}{\lambda_2})^{\xi} \cdot (\frac{1}{\varepsilon})^d \tag{69}$$

or, taking the algorithm on both sides,

$$d = 1 + \frac{\log_2 \lambda_1}{\log_2 \frac{1}{\lambda_2}} . \tag{70}$$

This formula holds for two-dimensional maps. It may be generalized, however, for higher dimensionality. We call d the "Lyapunov" dimension; under a number of circumstances it coincides with the information dimension.

Application: The two-dimensional "Horse-shoe" attractor

Starting from a rectangle ABCD (Fig. 8) we stretch it along the x-direction by a factor 2 and squeeze it along the y-direction by a factor of 2η with $\eta > 1$. We thus get the rectangle $A_1B_1C_1D_1$; we fold it as a horse-shoe and accommodate it within the initial rectangle as shown in Fig.8. The second iteration is shown in Fig. 9. An one-dimensional cross-section (Fig. 10) gives us a Cantor structure. The Lyapunov numbers are $\lambda_1 = 2$ and $\lambda_2 = 1/2\eta$ and the Lyapunov exponents are $\ln 2$ and $\ln(1/2\eta)$, respectively. Accordingly the fractal dimensionality of the Horse-shoe map is equal to

$$d = 1 + \frac{\ln 2}{\ln(2\eta)} = 1 + \frac{\log_2 2}{\log_2 2\eta} = 1 + \frac{1}{\log_2 2\eta} = 1 + \frac{1}{1 + \log_2 \eta} . \tag{71}$$

6.3. The concept of metric (Kolmogorov-Sinai) entropy and its relation to information dimension

So far we have been investigating the amount of information acquired by an observer in making a single, isolated measurement of a dynamical system posses-

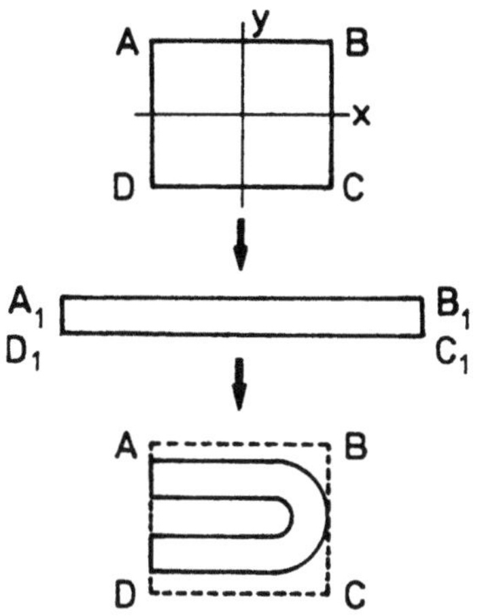

Fig. 8 The first iteration of the horse-shoe map

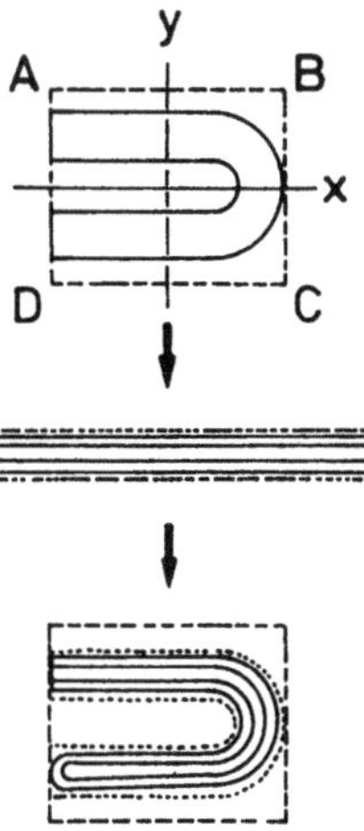

Fig. 9 The second iteration of the horse-shoe map

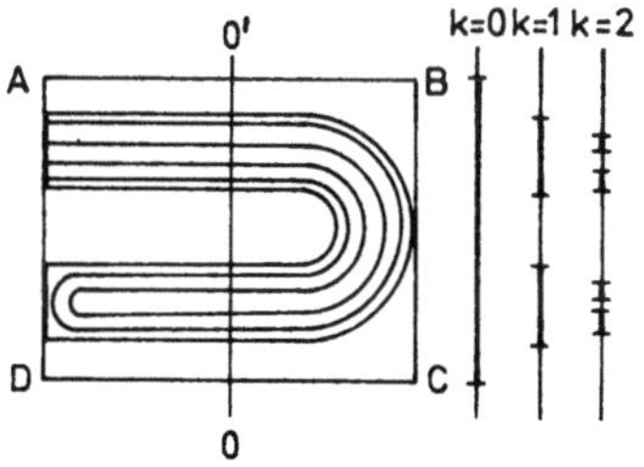

Fig. 10 The Cantor set structure of the one-dimensional cut of the horse-shoe map

sing a strange attractor or, in turn, the information an observer needs in order to define unambiguously one point on the attractor. We now turn to the "dynamical" problem. We assume that the attractor-possessing system "unfolds" before an observer, and we wonder about the new information the observer acquires by making successive measurements in time, that is, by receiving, from the system, a behavioural repertoire consisting of a discrete time series of "pulses". Obviously, if the attractor is a stable steady state, once the measurement process fixes that state, no further measurements are needed: no new information is to be acquired from a static object.

Likewise, if the attractor is a limit cycle, once the transients associated with the closing of one period are over, further observations merely reproduce redundant information. A similar thing happens to any other periodic attractor, e.g. a (rational) 2-torus. Things change dramatically, however, if the attractor is aperiodic (strange). In such a case the system is unpredictable and the metric entropy presently discussed provides an upper bound on the information acquisition rate from such an evolving dynamical system. Then, one immediately expects, that before the first "accumulation point" of the control parameter(s) of an attractor-possessing system (where the repertoire of the system consists only of stable periodic trajectories), the metric entropy should be zero and should start "taking-off" just after the appearance of the first aperiodic orbit.

Let us assume that the observer receives a discrete time series which constitutes an ergodic Markov chain possessing n states σ_i (in the form, say, of n integer numbers), revealing themselves through sequential measurements. The measurements are performed on each of the n elements or the partition of the state space which supports the attractor. If the occurrence of a certain symbol/state depends stochastically on m preceding symbols/states, the process is of mth order. Our first question is: What is the average amount of information per symbol emitted by such a Markov source? Clearly, the answer to this question should provide an observer familiar with the past history of the source with a measure of the unpredictability of the sequences generated by the source.

We now calculate the entropy of an ergodic Markov source of order m possessing a number of discrete states n. The total number of state sequences one may construct of length m out of n states is n^m. [Alternatively, if we consider each sequence S_m $(\sigma_1, \sigma_2, \ldots, \sigma_m)$ as an m digit base-n fraction, and arrange all those sequences as points in the interval $(0 \div 1)$, then two neighbouring numbers on the unit interval will have the distance n^{-m}, and there will be $1 - n^{-m}$ of them.]

Now each m-length sequence forms a "state" S_m in our m^{th} order Markov chain. The amount of information obtained when the next state is σ_{m+1}, given that the chain is now at the state S_m, is

$$I = -\log_2 P\left(\frac{\sigma_{m+1}}{\sigma_1, \sigma_2, \ldots, \sigma_m}\right) = -\log_2 P(\sigma_{m+1}/S_m) \ , \tag{72}$$

where $P(\sigma_{m+1}/S_m)$ is the conditional probability that, given the "state" S_m, the next $(m+1)^{th}$ symbol will be σ_{m+1}.

To obtain the entropy of the source we have to average the above expression over all possible transitions from S_m to σ_{m+1}, and over all possible n^m "states" S_m. Let $P(S_m)$ be the probability that the particular m-sequence takes place ($\sum_{m=1}^{n^m} P(S_m) = 1$). Then the average amount of new information per symbol σ acquired by the observer is

$$\Delta I_m = - \sum_{S_m=0}^{n^m} P(S_m) \sum_{\sigma_{m+1}=0}^{n-1} P\left(\frac{\sigma_{m+1}}{S_m}\right) \log_2 P\left(\frac{\sigma_{m+1}}{S_m}\right). \tag{73}$$

But by Bayes' rule,

$$P(S_m, \sigma_{m+1}) = P(S_m) \ P\left(\frac{\sigma_{m+1}}{S_m}\right), \tag{74}$$

where $P(S_m, \sigma_{m+1})$ is the joint probability for S_m and σ_{m+1} to occur successively. So we have

$$\Delta I_m = I_{m+1} - I_m \ , \tag{75}$$

where

$$I_m = - \sum_{S_m=0}^{n^m} P(S_m) \log_2 P(S_m) \ .$$

For an ergodic attractor, ΔI_m is then the average amount of new information gained by the observer as the trajectory of the attractor moves from "box" to "box" in the partitioned state space. It is essentially the (exponential) rate at which nearby trajectories diverge locally which is responsible for the generation of new information with each successive measurement.

Taking the limit as m goes to infinity we get the expression

$$\frac{\Delta I}{\Delta t} = \lim_{m \to \infty} \left(\frac{I_m}{m \Delta t}\right) \text{ bits/sec} \tag{76}$$

as the rate of information (bits/sec) generated by the unfolding attractor. The metric entropy is defined from the above expression if we now take the maximum value over all possible partitions ß of the state space, namely,

$$\text{metric entropy} = h_\mu = \sup_{(\beta)} \left(\lim_{\substack{m \to \infty \\ \Delta t \to 0}} \frac{I_m}{m \Delta t} \right) . \tag{77}$$

The metric entropy is then the upper limit (it assumes a perfect observer) of the average amount of new (not implicit in the initial conditions) information gained per second from the dynamical evolution of a strange attractor. As the window of resolution becomes narrower and narrower with increasing iterations, the above rate should increase since now small-scale random (thermal) fluctuations inherent to any large scale system are not smeared out and are bound to play a macroscopic role.

A practical way of achieving a rough calculation of the Kolmogorov-Sinai entropy consists in finding the correspondence between a discrete map iterated beyond the chaotic limit and a Markov chain: Take the example (Feistel (1983)) of the familiar map

$$X_{t+1} = 4X_t(1 - \frac{X_t}{\xi}) , \tag{78}$$

where ξ is an integer denoting the partition on the interval. For $\xi = 2$ we have a partition between two symbols A and B, say, on the interval 2 and for $\xi = 4$ we have a partition between four symbols, say, A, B, C, D (on the interval 4), (for $\xi > 4$ the partition is non-Markovian, see Feistel (1983)).

From geometrical considerations it is quite easy to calculate the elements of the transition probability matrix of the Markov chain involved. So, for $\xi = 2$ for example the portion of the A = 1 on the horizontal axis, which is projected as A to the vertical axis, equals $P_{AA} = 1 - (1/\sqrt{2})$; consequently $P_{AB} = 1/\sqrt{2}$; likewise $P_{BA} = 1 - (1/\sqrt{2})$ and $P_{BB} = 1/\sqrt{2}$. In case $\xi = 4$, the transmission matrix is as follows:

$P(\sigma_{t+1}/\sigma_t)$	A	B	C	D
A	α	0	0	α
B	β	0	0	β
C	γ	0	0	γ
D	0	1	1	0

where $\alpha = 2 - \sqrt{3}$, $\beta = \sqrt{3} - \sqrt{2}$ and $\gamma = \sqrt{2} - 1$.

From the elements of the transition probability matrix one can calculate the probabilities P(A), P(B) or P(A), P(B), P(C), P(D) from n - 1 of the linear relationships

$$u_K = \sum_{j=1}^{n} u_j P_{jK} \qquad (79)$$

and the normalizing relation

$$\sum_{K=1}^{n} u_K = 1 \qquad (80)$$

where $n = 2$ or 4.

So the entropy of the Markov chain involved equals

$$S = \sum_{K=1}^{n} u_K S_K = - \sum_{K=1}^{n} \sum_{i=1}^{n_K} u_K P_{Ki} \log_2 P_{Ki} \text{ bits,} \qquad (81)$$

where (for each case) n_K is the number of states which can be reached in a single hop from state K.

This concludes the section on the parameters characterizing a strange attractor. We have investigated them in some detail since all of them - information dimension D_I, spectrum of Lyapunov exponents $[\lambda_i]$, and metric entropy h_μ - are centred around two basic concepts:

a) The compressibility displayed by a strange attractor, that is, the degree of reduction of the numbers of degrees of freedom achieved by the fact that the flow maps onto itself, thereby forming a compact ergodic set in a state space with much higher dimensionality.
b) The concept of information generation, which makes (in finite time) the system independent of the initial conditions.

The investigation above is crucial for the main theme of this paper: namely the search of mechanisms ensuring the formation of collective abstract properties, thereby allowing the formation of hierarchical systems and the possibility of mutual simulation between them. Indeed compressibility is the sine qua non prerequisite for the formation of such collective properties via the formation of long-range cross-correlations among pertinent macrovariables. The speculation is that biological systems possessing "rhythmus" with chaotic properties, i.e. strange attractors, might be instrumental in performing reliable information processing in some "economic" fashion. Accordingly we next come to the first "application".

7. The electrical activity of the brain - should it be chaotic?

Brain-like structures have evolved by performing signal processing, initially by minimizing "tracking errors" on a competitive basis. Such systems are highly complex and, at the same time, notoriously "disordered". The functional trace of the cerebral cortex of the (human) brain is a good example.

The electroencephalogram (EEG) appears particularly fragmented during the execution of mental tasks as well as during the recurrent episodes of rapid eye-movement (REM) sleep. A stochastically regular or a highly synchronized EEG, on the other hand, characterizes a drowsy (relaxing) or epileptic subject, respectively, and indicates - in both cases - a very incompetent information processor.

We suggest that such behavioural change-overs are produced via bifurcations which trigger the thalamocortical non-linear pacemaking oscillator to switch from an unstable limit cycle to a strange attractor regime or, more correctly, from intermittency to metastable [1] chaos and vice versa. Our analysis aims at showing that the EEG's characteristics are not accidental but inevitable and even necessary, and therefore functionally significant.

An information processor (analogue or digital) is a cognitive gadget that tracks and identifies the parameters of an unknown signal or "pattern", which is usually contaminated by the thermal (equilibrium) noise (white or coloured, additive or multiplicative). In order to accomplish this task, the processor has to perform three distinct operations in the following sequence:

a) To produce patterns from "within" a wide variety of (spatial-temporal) patterns (or "templates").
b) To cross-correlate (i.e. "compress") each of these patterns with the incoming one.
c) To select or filter out, on the basis of some pre-established "hypothesis-testing" or "consensus" criteria, the pattern which forms the greatest cross-correlation with the unknown signal or trigger. (The filtering is usually non-linear in order to create and enhance contrast, a fact which, by sharpening contours, makes recognition simpler. Selected groups or cerebral neurons - and the Xerox machine! - do just that.)

To track a signal, timing is of essence. (The simplest tracker in use in communication engineering practice is the phase-locked loop (PLL).) This means that the existence of self-sustained non-linear dissipative oscillators (i.e. elements displaying attracting behaviour) at the hardware level of the processor is a prerequisite for the cognitive operation.

---

[1] Since the stationarity of EEG in many cases does not exceed the order of ~1 sec it is questionable if the process has the time to settle completely on one attractor.

Functionally stable oscillators, in contradistinction to static (switching "on"-"off") devices, offer indeed a number of evolutionary advantages, namely, (a) time keeping, (b) dynamic information storage (dynamic memory)[1), and when triggered by very simple stimuli, possibly (c) an extremely broad spectrum of complex behavioural repertoires.

Finally, the oscillators must by necessity possess asymptotic stability: one cannot accomplish reception and cognition tasks - which, by involving phase locking or compressibility, are dissipative (i.e. irreversible) - via Hamiltonian (reversible) working subsystems; hence the universality of the so-called family of van der Pol oscillators in communication engineering.

Parsimony - which undoubtly possesses survival value - requires that the locally generated dynamical patterns-attractors of the processor should not always be "on". They rather should emerge upon request (i.e. upon triggering from externally impinging stimuli) from a set of available dynamical elements and some basic and rather simple recursive <u>rules</u> (algorithms of "maps") for combining those elements.

Below we present a sketch for a dynamical model of a brain processor. Individual neuronal oscillators at the cerebral cortex constitute the above-mentioned set of dynamical elements. The thalamocortical oscillator, on the other hand, is the adaptive agent which performs two distinct operations:

a) It provides pacemaking activity, resulting in the formation of internal, synchronized, or coherent (spatial-temporal) neuronal patterns, each of which stands for an attractor. By making such neuronal groups coherent, the pacemaker helps them to elevate themselves above the ambient thermal noise level and also to distinguish themselves from coexisting neighbouring neuronal formations, within brief time intervals.
b) It generates the <u>recursive rules</u> governing the sequential appearance of these coherent patterns-attractors on a time-divison-multiplexing scheme.

7.1. Experimental data from EEG research

In recent years, a very great number of studies (see Adey (1968, 1970, 1972), Elul (1966a,b, 1968, 1969a,b, 1972), Verzeano (1972), Verzeano et al. (1970), and references therein) has been devoted to the electrical activity of the (human) brain - activity both spontaneous and evoked - and hypotheses have been advanced concerning the mechanisms and the functional significance (if any) of the EEG. It must be stressed from the outset that the brain is <u>not</u> a conventional signal processor: due to its enormous redundancy, different departments of the whole system are permeated with "noises" of different kinds, performing

---

[1)] Dynamic memory refers to storage of an algorithm from which patterns can be produced rather than to storage of the bare patterns themselves.

overlapping triggering functions at different times (Nicolis and Benrubi (1976)) and having varying degrees of "relevance" in catalyzing specific metabolic processes.

At any rate, the experimenter's opinions seem to converge nowadays in a number of fundamentals:

a) In pronouncing the EEG, the total sum of elemental "slow" neuronal self-sustained oscillatory activities (graded, sub-threshold post-synaptic membrane potentials) crowded in rather small volumes of (cerebral) cortial tissue just underneath the specific recording electrode (scalp lead).

b) In reading spikes travelling along individual neuronal axes as irrelevant for the generation and form of the EEG, apart from their causing non-linear coupling among the elemental generators (via pre-synaptic excitation and neurotransmitter/hormonal discharge into the synaptic cleft).

c) In finding no consistent phase relationships between each of the elemental neuronal oscillators and the gross EEG. In terms of amplitude, the intracellular oscillators are in the range of 50 mV; the EEG in the range of 100 µV. In terms of spectrum, they are more or less identical (narrow-band noise in the interval $\sim 0.3 \div 45$ Hz). From the point of view of statistical characteristics, the amplitude probability density function of each elementary oscillator is non-Gaussian (Elul (1969)), while the amplitude p.d.f. for the EEG is Gaussian in the "idle" regime, but becomes more or less skewed during REM sleep and during the execution of (controlled) mental tasks.

So much for the "hard" established data. The following includes what are essentially inferences from the above evidence and further experimental results, still not universally accepted.

A fundamental question is whether the EEG represents activity of synchronized (mutually phase-locked) groups of neuronal oscillators or whether it is produced via summation of statistically unrelated phasors. The experimental probing of the two alternatives has been quite extensive. The current view (Elul (1972)) compromises the two extremes und supports the likelihood that the EEG may be produced via sequential and intermittent synchronization of successive, relatively small groups of cortial neurons (not necessarily from the same cerebral area), each of which forms a small (1% - 10%) subset of the entire neuronal population. Moreover, we are capable of recording activity only for that particular subgroup which happens to be synchronized at the given moment and consequently attains detectable voltage levels (above ~ 5 µV).

Further, it is indicated that this sequential phase-locking - or more generally, the establishment of long-range space-time coherence among the members of the individual neuronal subgroups (Adey (1972)) - is mediated via specific thalamic pacemaking groups of neurons (relaying sensory cues to the cortex). The existence of thalamocortical and corticothalamic feedback-feedforward loops (the latter ending on non-specific thalamic nuclei, whose functional role is still elusive) has been experimentally traced and mapped in detail (Verzeano (1972)). These pacemaking loops are further considered responsible for the scanning or the sampling process, which allows the sequential formation and breaking of phase-ordered relationships among the neuron members of the individual subsets.

Such a mechanism provides a time-division-multiplexing of information to be processed at the neocortical level. So, although the EEG appears as a continuous stochastic process, it may in fact be considered as a time sequence of discretely evoked potentials, each possessing an average duration of $\sim 100 \div 300$ msec. Synchronization then aims at enhancing - one at a time - patterns of specific groups of cortical neurons, which then become able to process and store, on a collective basis, specific sensory modalities via (neuronal) pre-post-synaptic membrane $\rightleftarrows$ cytoplasmatic $\rightleftarrows$ nucleic, cascade loops.

7.2. The model

We emulate now the "thalamocortical pacemaker" as a high amplitude dissipative self-sustained non-linear (relaxation) oscillator, which in the absence of any environmental input, is intermittently "free running" on an unstable limit cycle with a fundamental ("sampling") frequency ~ 10 Hz (the "a-rhythm"). This means that the oscillator has an internal bias, which is responsible for this instability.

The way such a strong oscillator can synchronize groups of "smaller" oscillators has been sketched (Nicolis et al. (1977)). If our conjecture is relevant, we should expect the corresponding EEG to exhibit some pseudo-periodicity but also randomness, in the sense that successive amplitude segments of this activity (each segment corresponding to a given stationary state/pattern at the cortical level, formed by the vector sum of the amplitude of phase-locked neuronal phasors of the pertinent cortical subset) should show no statistical correlation to each other. (The optimal sampling of a random process is at random time intervals: the sampled modalities whose embodiments are the sets of synchronized neurons should thus not overlap.)

This seems to be indeed the case: the amplitude p.d.f. of such an EEG follows a normal distribution, as one could expect from the central limit theorem (or the adherence to the law of averages)[1].

---

[1] There seems to be a contradiction here: if the postulated mechanism is indeed intermittency I, the amplitude p.d.f. should not be Gaussian but hyperbolic. However, one should not forget that the processor divides its time not between <u>one</u> limit cycle and irregular motion, but among <u>many</u> indistinguishable limit cycle modalities. The experimentally derived p.d.f. therefore represents an ensemble average of individual hyperbolic distributions and it may well be Gaussian.

Consider now what may happen in the regime of REM sleep or the performance of a mental task: information is pumped along the ascending branch of the reticular formation system or through other viaducts of the peripheral nervous system, carrying, from the inside or from the outside, specific sensory inputs to be identified by the cortex. Our thalamocortical oscillator is now under the spell of a fluctuating input, and beyond specific (threshold) values of some control parameter, a cascade of bifurcations may set in, as a result of which the intermittent limit cycle is deleted, and the oscillator follows now the part of a strange attractor. In other words, it becomes metastably chaotic; its scanning manifestations on the cortical neuronal subgroups turn to a "spasmodic" and non-periodic oscillation.

This hypothesis is corroborated by long-ago established experimental evidence (Moruzzi and Magoun(1949)).Increasing arousal of the ascending reticular formation branch leads to a polarization of the pacemaking-specific thalamic nuclei, thereby interrupting their sampling function for time intervals concomitant with the degree of arousal, i.e. the intensity of stimulation and therefore the rate of information transfer from the environment. So, under excitation, the sampling of the cortex from the pacemaking thalamic nuclei switches from intermittently periodic to sequential. We witness the establishment of a Markov chain with different corresponding probabilities u_i per stationary synchronized state i (internal pattern) - produced from a transitional probability matrix P_{ij} between successive states/patterns. So,and this is the crucial point, the amount of "holding" time that the cortical processor spends at the specific modality i depends on the amount of "disruption" time caused in the scanning pacemaker at i by the information input; this time usually, but not always, increases with the rate or the frequency of the incoming trigger.

Evidence supporting this view comes from analyses of the amplitude probability distribution of EEG indicating changes from Gaussian to skewed forms during the performance of mental tasks (Verzeano et al. (1970)). We are afraid, however, that the interpretation put forth in the above reference, that the skewness in the p.d.f. reflects some "increasing degree of cooperativity among the neuronal generators" is irrelevant: changes of coupling among neuronal oscillators embodying a pattern may come from different causes; they result only in changing (enhancing or deteriorating) the degree of coherence of the specific pattern to which they contribute but, of course, this has nothing to do with the algorithm of time-division-multiplexing or the policy of switching among patterns.

We interpret the observed skewness as indicating a statistical linear correlation ($P_{ij} \neq 0$) or a Markovian characteristic between the successive amplitude EEG samples, i.e. between successive synchronized neuronal states, each coding for a single sensory modality. Under such linear interdependence, the central limit theorem of course does not hold.

Suppose finally that the degree of arousal of the reticular formation ascending branch, or the average rate of information pumping, increases still further. We make the conjecture that in such cases the scanning process of cortical subgroups either gets extinguished (the oscillator gets "quenched"), or the scanning rate increases concomitantly with the degree of arousal. In the limit we may consider that all the subgroups of cortical neurons are synchronized at the same time, a symptom of epileptic convulsions. We do not possess reliable experimental data, and certainly a proper theory is still lacking. Nevertheless, in a recent computing simulation (May (1981)) hints have been offered that when the coupling between the environment and a limit cycle oscillator, already in the chaotic regime, becomes "supercritical", one may witness a regression from the chaotic regime back to a limit cycle regime. (The same regression can be achieved by external noise; see Nicolis and Tsuda (1985).)

The hint is relevant to the possible behaviour of our thalamocortical pacemaker, which, under strenuous external (photic and acoustic) triggering at its fundamental frequency or one of its harmonics, may switch back from the chaotic to a strictly entrained regime with such a high amplitude of oscillations that it can virtually synchronize large numbers of cortical neurons simultaneously.

To sum up, we have tried to appreciate why a cognitive system may be chaotic in order to perform effective signal processing. The answer can be offered in the domains of (a) time and (b) frequency.

a) By creating a Markovian time-division multiplexing, the system allows separation of sensory modalities-attractors and relegates for each of these modalities a time-processing interval commensurate with the rate of sensory input.
b) "Turbulent chaos", (non-equilibrium, low-dimensional noise) as a modus operandi of the thalamocortical pacemaker under mild excitation, contains a broad spectrum of temporal (and spatial) frequencies. Thus it can constrain "patches" of post-synaptic functional areas in the cortex and create coherent patterns which can match a wide variety of incoming spatial-temporal patterns.

Specifically, cognition is manifested at the cortex as a result of a matching process between pairs of spatial-temporal patterns, each containing a great number of elemental units (neurons). In each pair, one pattern (the same for all pairs) is the unknown information; it is embodied in incoming triggers, coded either in sequences of pulses from the peripheral nervous system, or, if it comes from other areas of the central nervous system, encoded in strings of of macromolecular (neurotransmitter/hormonal) releases from pre-synaptic endings.

The other pattern of the pair is one of the patterns - attractors created by the processor; it constitutes a prestored spatial-temporal "mosaic" embodied in a set of partly synchronized post-synaptic membrane potentials or a spatial-temporal pattern of post-synaptic membrane receptors.

The coupling or cross-correlation between the above two patterns of each pair takes place dynamically via energy exchanges between equal or neighbouring frequency pairs (ω_i, ω_j) shared by both spectra, and is estimated quantitatively by the coherence function

$$\mathrm{coh}_{ij}(\omega) = \frac{\varphi_{ij}(\omega)}{\sqrt{\varphi_i(\omega)\,\varphi_j(\omega)}} , \tag{82}$$

where $\varphi_{ij}(\omega)$ is the cross-spectrum and $\varphi_i(\omega)$, $\varphi_j(\omega)$ are the auto-spectra of the two matched patterns.

The result of the cross-correlation in phase and amplitude determines the "degree of cognition" between the incoming and the preset of the unknown and the expected patterns. Which mechanism in the brain decides which coherence function is the predominant? It appears that again the reticular formation might be involved: some plausible ways of non-linear filtering execution have been discussed in Kilmer and McCulloch (1969).

In this application we have attributed the phase-locking between individual neuronal oscillators (and therefore the formation of attractors in the cortex) only to the thalamocortical pacemaking activity. Undoubtedly there are other, perhaps equally important, very complicated mechanisms of extending long-range spatial-temporal coherence among cortical neurons, mechanisms which have received, exceptions granted (Adey (1972)), very little attention. We just mention the possible role played in this respect by the cortical intercellular electrolyte.

The intercellular fluid appears to contain extensive hydrated networks of complex membrane-bound macromolecules (mucopolysaccarides and mucoproteins). The removal of calcium allows those moleculs to bind with water and is concomitant with their spreading in a loose hydrated network. Restoration of calcium reverses this process, unbinding water and contracting the molecular net, up to five orders of magnitude in volume.

The existence of very weak EM fields seems to influence strongly Ca_{++} movement in the intercellular electrolyte, as well as the impedance of the fluid. We thus may envisage new ways of establishing and controlling long-range coherence in the brain and thereby influencing the formation of attractors in the cortex: random environmental EM fields. These fields, standing as external noise, may provide the mechanism by switching from attractor to attractor.

7.3. Deriving the existence of an attractor and its fractal dimensionality from one-dimensional time series

Consider a (discrete) time series X(t) standing for sequential measurements we perform on a certain variable of a dynamical phenomenon. We are interested, with the sole knowledge of this time series, in investigating the nature of the phenomenon involved, namely whether it is random noise or an attractor, and furthermore we are interested in estimating the minimum possible number of independent variables whose interplay in time would essentially account for the (multidimensional) phenomenon involved.

To this end Grassberger and Procaccia (1983) suggested an elegant way, which, in broad lines, is as follows: Let $d\underline{X}(t)/dt = \underline{f}(\underline{X}\,;\mu)$ be a system of N coupled non-linear ordinary differential equations standing for the mathematical model of the physical system under investigation. N is still unknown. The trajectory of the system is described in an N-dimensional state-space with coordinates $X_1, X_2, \ldots, X_N$.

Under some conditions the above system can be reduced to an Nth order non-linear ordinary differential equation with respect to X, where X is the variable whose time series we measure. Then the state-space configuration with coordinates X(t), dX/dt, ..., $d^{N-1}X/dt^{N-1}$, is topologically equivalent with the original one - and so is the state-space configuration with coordinates X(t), $X(t+\tau)$, $X(t+2\tau)$, ..., $X[t+(N-1)\tau]$, where τ is an arbitrarily taken time lag. (In practice, in order to get good "phase portraits" it is advisable to choose τ not too small.)

So, from the knowledge of the discrete time series X(t) we may construct the trajectory of the phenomenon. In a chaotic regime (corresponding to a strange attractor) the positions of two points ($\underline{X}_\kappa$, $\underline{X}_\lambda$, say) in the trajectory - sufficiently apart in time - are, uncorrelated by definition. Nevertheless, once we are on the attractor there exists some spatial correlation between $\underline{X}_\kappa$ and $\underline{X}_\lambda$.

For a given N, say N_1, we consider m pairs of points $\underline{X}_\kappa$, $\underline{X}_\lambda$ and investigate their distance $|\underline{X}_\kappa - \underline{X}_\lambda|$ for all $\kappa, \lambda \in (1, \ldots, m)$. We choose an arbitrary distance r and investigate the difference $|\underline{X}_\kappa - \underline{X}_\lambda| - r$. If this is positive, we write 1, and if it is negative, we write 0. We thus get one point of the correlation function

$$C(r) = \lim_{m\to\infty} \frac{1}{m^2} \sum_{\kappa}^{m} \sum_{\lambda}^{m} H\left[r - |\underline{X}_\kappa - \underline{X}_\lambda|\right], \qquad (83)$$

where H is the step function. We repeat this for a great number of pairs (say $m \sim 10^4$). The number of pairs κ, λ whose distance is below r is, on the other hand, proportional to r^d, where d is the dimensionality of the attractor.

So

$$C(r) \sim r^d \tag{84}$$

and

$$\log C(r) = d \log r \, . \tag{85}$$

We repeat the above procedure for different values of N, N_1, N_2, ..., and plot logC(r) vs. logr for each individual N_i. We get a group of graphs sketchily shown in Fig. 11. From these curves we next plot d versus N (Fig. 12). If d is linearly increasing with N, then the phenomenon involved is obviously <u>stochastic</u> (random noise) with a great number of degrees of freedom. If, however, the curve d = f(N) levels off and saturates beyond a certain value $d_c = N_c$, this is an indication that we have a strange attractor with a <u>minimum</u> number of degrees of freedom d_c. In the case of the EEG some preli-

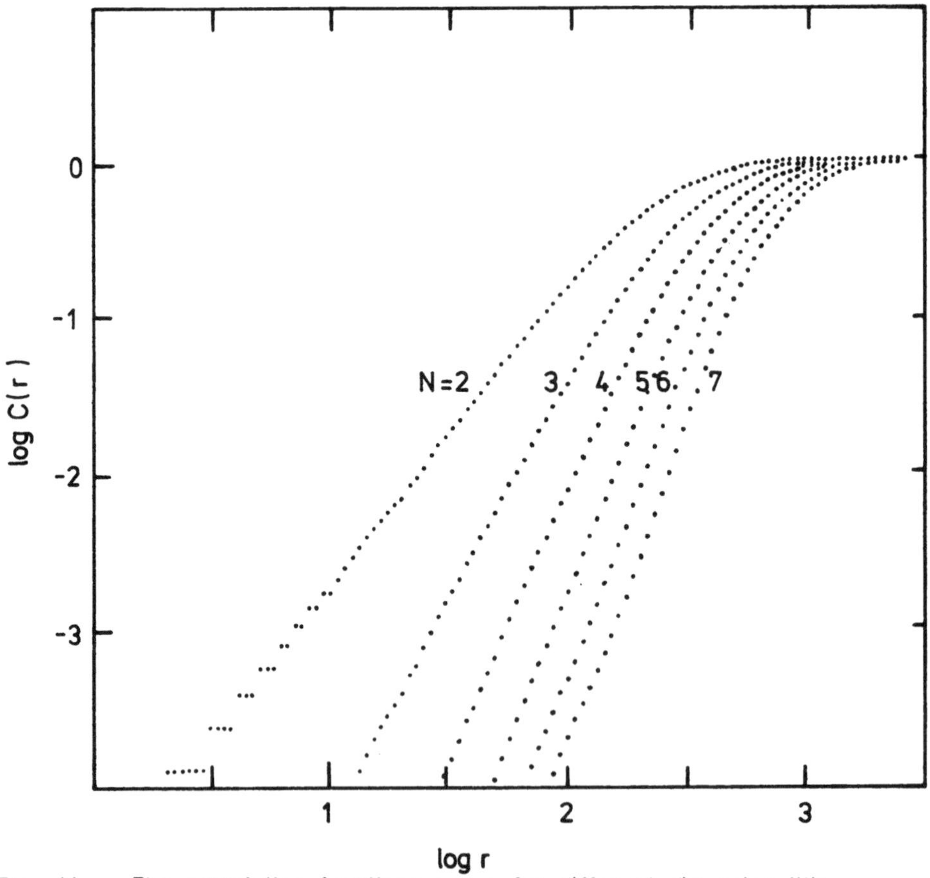

Fig. 11 The correlation functions vs. r for different dimensionalities Ni of state space

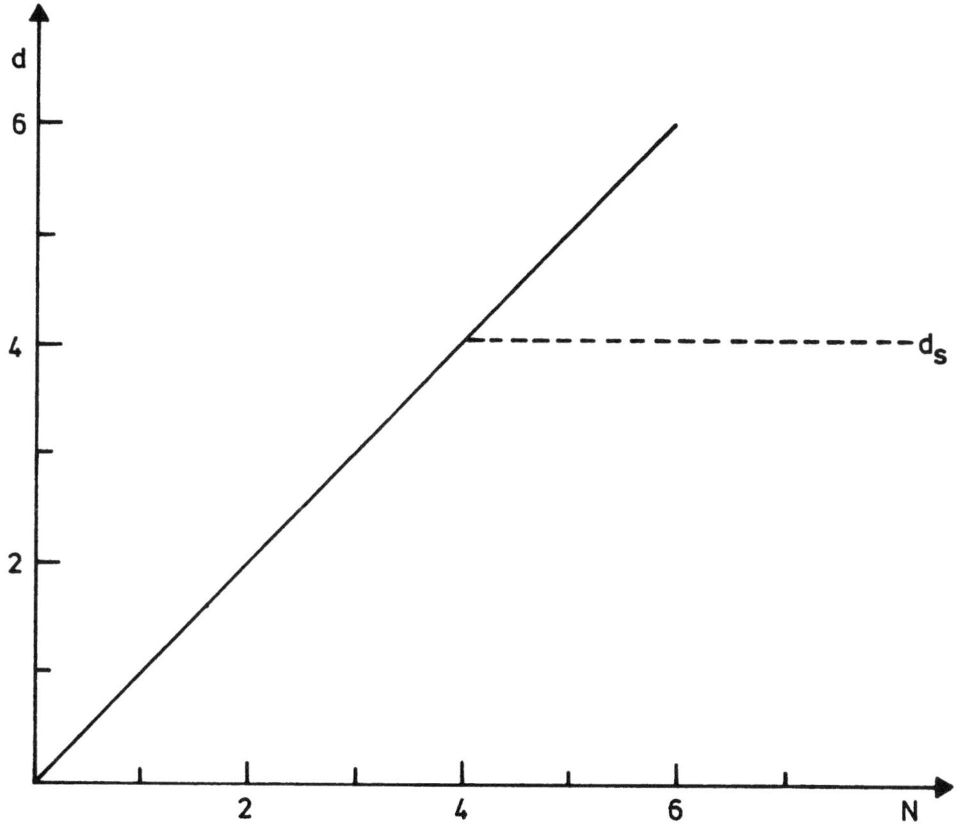

Fig. 12 Fractal dimensionality vs. state space dimensionality. Saturation implies the existence of an attractor

minary investigations by A. Babloyanz and C. Nicolis on samples taken from healthy human adults during the dreaming [(REM sleep)] regime show a value $d_c \sim 4 \pm 0.1$. In the awake state it appears that d_c is much higher (of the order of ~ 7).

7.4. What is a cognitive channel?

The human mind possesses the unique capability of "mapping" the external (as well as part of the organism's internal) world - that is, it "compresses" long and complex strings of impinging environmental stimuli ("observations") - and then uses these "minimal length algorithms" in order to simulate physical phenomena - thereby revealing the "laws of nature".

We may theorize that this process of "self"-organization and category formation is implemented via a set of coexisting (strange) attractors in the cognizant apparatus - each of which attracts (and therefore compresses) whole

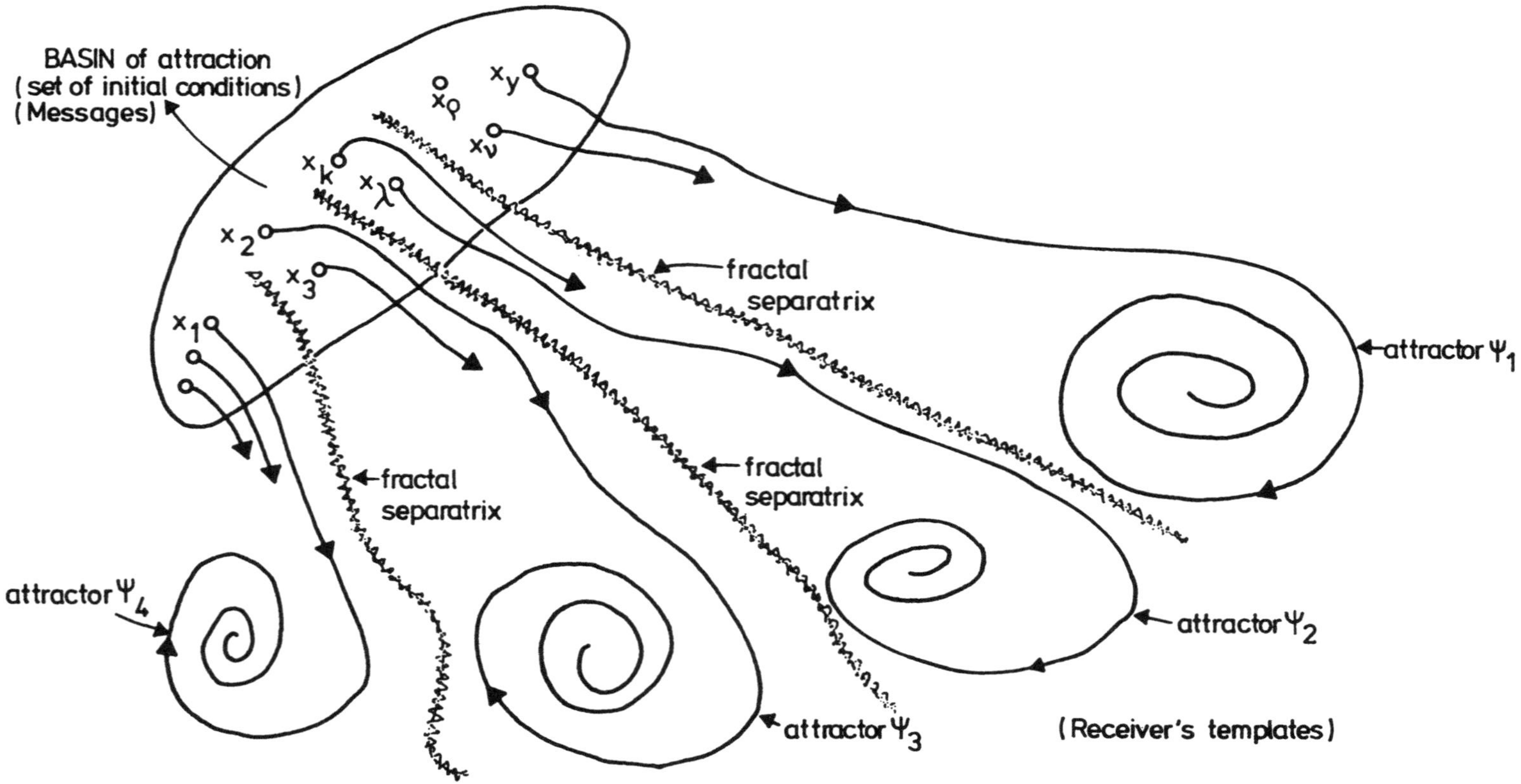

Fig. 13 Sketch of a cognitive channel. The x's and y's belong to different hierarchical levels

subsets of "initial conditions", the total sum of which constitutes the set of external stimuli (Fig. 13).

This set of initial conditions forms the "Basin" of the attractors, and the processes of partition and category formation in the mind involves the topology of the separatrices among the individual subsets of the Basin.

Let us then try to sketch the dynamical equivalent of an information channel mediating the mapping between a set of messages of the "transmitter" and a set of "regimes" of the "receiving" or rather the cognitive apparatus - playing the role of a memory bank. (Let us recall again that an information channel according to Shannon's "static" information theory provides a one-to-one mapping between the transmitter and the receiver and is in no way associated with the act of "cognition". The sole role of a classical information channel is the mere copying of the transmitted patterns, as they are, at the receiving end with the greatest possible reliability.)

We propose that the set of messages should be identified as a set of initial conditions or the Basin of the attractors in an N-dimensional state space, where N stands for the number of variables mediating the dynamical phenomenon under consideration.

The receiver, or, more generally, the cognitive apparatus is identified as the set of attractors of much lower dimensionality, embedded as invariant compact subsets in the same N-dimensional state space.

Falling into the attractor via (short or long) transients in state space corresponds to the act of progressive signal abstraction (or compression). The issue is: what about the dynamical isomorphism of the "channel"? or, how is the set of initial conditions, that is, the messages, partitioned among different categories, that is, among more than one coexisting attractor (steady states, limit cycles, tori or strange attractors)? One has to realize here that the mapping between messages and attractors (or "templates") is not one-to-one: whole subsets (of initial conditions) of the Basin converge towards one and only one attractor. Alternatively some points of the Basin may undergo a perpetual (or too long) transient trajectory without converging anywhere, or they may even be repelled to infinity. How can we say which subset of the Basin will land into which attractor?

The problem in this general form is, of course, of formidable difficulty. In a very elementary case (two simple attractors, namely two steady states) the issue has been recently addressed. One considers a two-dimensional non-invertible map (the "channel") and finds a fractal separatrix between the two subsets of the Basin (Grebogi et al. (1983)).

The essential outcome of the above work is that the precision with which a given initial condition is known may play an extraordinary sensitive role on the predictability of the final attractor upon which this initial state will eventually land. This means that an extraordinarily high accuracy is sometimes necessary in our measurements for the reliable prediction of the pertinent attractor ("compressor") into which the observed phenomenon will be categorized: clearly then, in our model the set of attractors plays the role of dynamically stored patterns in the brain against which whole sets of externally impinging stimuli have to be "convoluted". This process may not be accomplished in one single hierarchical step; in a second step the attractors (if numerous at the lower level) will play the role of the members of a hyper-Basin towards a new hierarchy of fewer hyper-attractors and so on.

Knowing the state-space dimensionality of the impinging messages and the (fractal) dimensionalities of the corresponding attractors one can evaluate in each case the degree of (average) compressibility which a given N-dimensional stimulus is subjected to.

Furthermore, investigating the topology of the separatrices of the Basin one can calculate the uncertainty (or the information gained thereof) associated with the category to which the given stimulus belongs. Together, the entropy of partitioning and the degree of compressibility give the two essential macro-parameters characterizing the cognitive process at a given hierarchical level[1).]

Once the mapping (that is, the "linguistic scheme" adopted by the individual) is known, the two above parameters can on principle be calculated.

Already in the Introduction we have appreciated the fact that the percentage of compressible sequences in a string, say of 0's and 1's of length N, is extremely small. Considering the 2^N a priori equiprobable sequences of length N (the members of the Basin) the number of compressible sequences to, say, K bits is just

$$\Sigma = 2^1 + 2^2 + \ldots + 2^{N-K-1} = 2^{N-K} - 2 \tag{86}$$

and the percentage of compressible sequences is

$$\eta = \frac{\Sigma}{2^N} \sim 2^{-K} \quad . \tag{87}$$

So, although on principle there is a great amount of attractors at each cognitive level, the percentage of those which really deserve the name, that is, possess an information dimensionality $c = N - K$ bits ($K \sim 0(N)$), is just $\eta = 2^{-K}$, which is insignificantly small.

The moral is that the human mind can simulate a very small subset of natural phenomena. Their overwhelming majority literally "passes through" our heads undetected.

Nevertheless, how can it be explained that our templates-attractors really

---

1) According to our model, the EEG is the pacemaking agent which permutes the attractors on a time division multiplexing basis.

have the set of externally impinging stimuli as Basin? To accept this one has, a priori, to digest the hypothesis that there exists a "homology" between some patterns of the external world and the categories of the software of the human brain.

But if the human brain is just an unsuspected evolutionary byproduct of the "laws of nature", how is it that its software can simulate the very laws which, acting uppon matter, triggered its formation?

7.5. An example of a fractal Basin boundary (separatrix)

Grebogi et al. (1983a) give the following example of a two-dimensional non-invertible map:

$$\theta_{n+1} = 2\theta_n (\mathrm{mod}\ 2\pi)$$
$$z_{n+1} = \lambda z_n + \cos\theta_n \ , \quad 2 > \lambda > 1, \quad 0 \leq \theta \leq 2\pi \ . \tag{88}$$

The Jacobian matrix of this map has eigenvalues 2 and λ. Since both exceed 1, there are no attractors with finite z. In fact, all initial conditions will produce orbits which will asymptotically go either to $z = +\infty$ or $z = -\infty$ as $n \to \infty$. So we consider $z = +\infty$ and $z = -\infty$ as two coexisting attractors and ask what their Basin boundary is. This means: what is $f(\theta)$ like such that initial conditions in $z > f(\theta)$ asymptotically lead the orbits to $z = +\infty$, while those in $z < f(\theta)$ lead the orbits to $z = -\infty$.

Grebogi et al., starting from the above map, find the expression

$$z = f(\theta) = - \sum_{\kappa=1}^{\infty} \lambda^{-(\kappa+1)} \cos(2^{\kappa}\theta) \ . \tag{89}$$

Since $\lambda > 1$, the above sum converges.

Now investigate $df/d\theta$. From the above expression we get

$$\frac{df}{d\theta} = \frac{1}{2} \sum_{\kappa=1}^{\infty} \left(\frac{2}{\lambda}\right)^{\kappa+1} \sin(2^{\kappa}\theta) \ .$$

Since $\lambda < 2$, this sum diverges and so $f(\theta)$ is not differentiable. In fact, Grebogi et al. find that the separatrix $z = f(\theta)$ has an infinite length and a fractal dimensionality

$$d = 2 - \frac{\ln\lambda}{\ln 2} = 2 - \log_2\lambda \ . \tag{91}$$

A final important aspect of the information dimensionality of multiple existing attractors-"classifers" is the problem of their invariance under smooth coordinate transformations. In other words, we are interested in knowing whether the degree of abstraction or compressibility of the Basin of initial conditions possesses universality or not. We do not have a general answer to this question. Ott et al. (1984) have recently investigated the possible invariance of a number of "dimensions" of strange attractors and have found some specific coordinate transformations ensuring the invariance of the information dimensionality only. Clearly, much more work has to be conducted along this direction.

8. The problem of computational complexity

Although a strange attractor as an object of observation frustrates the (naive) observer, due to his practical inability to deduce reliable predictions from a series of measurements he performs on the system, a strange attractor as a cognitive device possessed by the observer plays exactly the opposite role, namely it gives rise to the compressibility of seemingly chaotic sequences of observed phenomena[1). It is perhaps the possession of such "cognitive devices" by the software of the brains of higher mammals (almost exclusively humans) which makes possible social cohesion, science, technology, and economic and cultural development in an evolving world. Since the most elemental and basic intellectual activity is computation, it is perhaps relevant to ask how humans compute and whether the possible existence of chaotic attractors in their brain function can be held responsible for the way decision making takes place. The prompt answer is, of course, that we do not have the slightest idea. Still we do suspect that the human way of computing cannot be only serial (sequential) but perhaps hierarchical also. Let us elaborate on this point by referring to what Bremermann developed some years ago as the physical limitation of serial computation; this is known as the "Bremermann's limit". (For an updated version, see Bremermann (1982)).

The idea is that any signal-processing activity involving information storage and retrieval should be passed on some "material" background (hardware) such as punched cards, tape recording, and so on. Suppose we have a mass m of a physical body which is used for this purpose of processing, with 100 per cent efficiency. The most natural way to store information which can subsequently be retrieved unambiguously is to "deposit" each bit on each energy level of the material body. How many bits (n) can be stored in this way in a mass m ? Obviously,

$$n = \frac{E_{max}}{\Delta E} , \qquad (92)$$

where E_{max} is the maximum amount of energy one can extract from the physical body, and ΔE is the resolution with which one energy level can be observed.

But $E_{max} = mc^2$, and from the uncertainty principle $\Delta E \cdot \Delta t \sim h$, where Δt is the oberservation time and h the Planck constant. So the maximum possible number of bits that our processor can manipulate in time Δt is

1) In the simple case of the Lorenz attractor, for example, the fractal dimensionality as calculated by Mori (1980) is $D_F \sim 2.06$. The compressibility then has a minimum value $N - D_F = 3 - 2.06 = 0.94$ bits. So any initial condition under the action of the attractor is compressed by almost 30 per cent.

$$n = m \frac{c^2}{h} \Delta t \quad \text{bits} \tag{93}$$

or

$$n_0 = \frac{c^2}{h} \quad \text{bits/gr.} \cdot \text{sec.} \tag{94}$$

This number is

$$n_0 \sim 2 \cdot 10^{47} \quad \text{bits/gr.} \cdot \text{sec.} \tag{95}$$

Suppose now that the whole physical universe were a perfect information processor that has been working on this job since the moment of the big bang. How many bits (n') could the universe have processed serially during its lifetime (so far)? The age of the universe is roughly $\tau \sim 10^{17}$ sec, and its total estimated mass $m \sim 10^{58}$ gr. Then

$$n' = 2 \cdot 10^{47} \cdot 10^{17} \cdot 10^{58} \sim 10^{122} \quad \text{bits.} \tag{96}$$

Is this number big or small? Well, in order to appreciate what humans do in that respect let us consider an information "box" with Λ inputs and r outputs and ask about the rate of information flow in such a box. We take as an example a big international airport where the inputs stands for say 1000 requests in dyadic form ("to land or not to land?") per day and the output with $r = 2$ stands for "Yes", "No". The traffic is handled by $\sim$ 20 men - air controllers. How much information flows into the box per day?

The total number of complexions is $w = r^{\Lambda}$, where $r = 2$ and $\Lambda = 2^{1000}$; the amount of information is then

$$\log_2 w = \log_2 r^{\Lambda} \tag{97}$$

or

$$\log_2 2^{2^{1000}} = 2^{1000} \sim 10^{300} \quad \text{bits.} \tag{98}$$

(Dividing this amount by the number of seconds in 24 hours multiplied by the number of air controllers does not help much!) How do humans (or machines designed by humans) handle in 24 hours an amount of information 180 orders of magnitude above the amount of information that a perfect serial computer possessing the size and the age of the universe can process. The answer is by "short cuts", that is, through instabilities giving rise to cascading bifurcations, beyond each of which the hierarchically arising new pattern(s) may provide the solutions ! Such patterns could take eons to emerge if one had to produce them serially. Bifurcations, then, and the emering higher hierarchical structures may provide the "deus ex machina" for solving complex problems by "intuition". Still, there are particular problems, the so-called NP-complete problems (NP standing for non-deterministic polynomial time), which stubbornly refuse even an algorithm for their solution[1)].

Take, for example, the classical problem of the "travelling salesman" . A salesman has to visit N cities, passing at least once each of them, and return to the point of departure. How should he schedule his visits in order to minimize the total distance? Regrettably, there is still no algorithm available for this problem. The salesman has to try before his departure all possible combinations which are N! or if N is large, $\sim N^N$ after applying Stirling's formula. We say that the number of calculations needed increases exponentially with the complexity of the problem. In tractable problems the number of calculations grows only as a polynomial function of the size of the problem. The difference between exponential and polynomial solutions in terms of computing time may be huge. For example, if N = 30, 3^N computations take $\sim$ 6.5. years of conventional digital computer time, but N^3 computations take just $\sim$ 0.027 seconds. If N = 60, 3^N computations take $\sim 10^{15}$ years, whereas N^3 computations take merely $\sim$ 0.216 seconds. Should the design of chaotic processors in the future alleviate the difficulty? Biological organisms most probably possess such processors (the enzymes and the cerebral cortex are good candidates). Still, we do not yet know how they work and so we cannot design artifacts mimicking or surpassing them.

1) "Analog computers" using physical bodies in the vicinity of "phase transitions" could provide the machinery needed here.

9. Application to cognitive processess: The "magic number seven" plus-minus two revisited

In a well-known collection of his essays in Cognitive Psychology George A.Miller describes in detail a number of experiments aiming at a determination of the limits (if any) of the human brain in processing information. He concludes that the "channel capacity" of human subjects does not exceed a few bits or that the number of categories of (one-dimensional) stimuli from which unambiguous judgement can be made are of the order of "Seven Plus or Minus Two". This "magic number" holds also, Miller found, for the number of random digits a person can correctly recall on a row and also the number of sentences that can be inserted inside a sentence in a natural language a human can read throughout without confusion. In this section we propose a dynamical model of information processing by a self-organizing system which is based on the possible use of strange attractors as cognitive devices. It comes as an amusing surprise to find that such a model can, among other things, reproduce the "magic number seven" plus-minus two and also its variance in a number of cases and provide a theoretical justification for them. This justification is based on the optimum length of a code which maximizes the dynamic storing capacity for the strings of digits constituting the set of external stimuli.

This provides a mechanism for the fact that the "human channel", which is so narrow and so noisy (for the order of just a few bits per second or a few bits per category) possesses the ability of squeezing or "compressing" practically an unlimited number of bits per symbol - thereby giving rise to a phenomenal memory.

9.1. Review of experimental results

Central amongst the aims of Cognitive Psychology is the evaluation of human performance during the execution of mental tasks. In order to measure this performance in terms of bits of information it is instrumental to emulate the subject as an information channel. To this end a well-defined finite set of alternative stimuli (input) is provided by the experimenter. Such stimuli may be strings of digits, letters, words, geometrical symbols, tones, pictures, etc. To each of these stimuli a definite a priori probability of occurrence is assigned (although in practice, that is as far as the subject is concerned, all stimuli are presented with equal probability). Then, a well-defined finite set of alternative responses (output) is selected (motor, vocal, reading, calculating, writing, etc.). Finally, the stimuli are presented to the subject in a randomized order at a steady rate in time and the subject's response to each stimulus is recorded.

From a long, statistically significant, sample of recorded data the measure of the transmitted information by the human channel can be evaluated as the input-output (stimulus-response) crosscorrelation. The "channel capacity" of the subject is the upper limit of the extent to which the subject can match his responses to the given set of stimuli. Instead of increasing the rate of stimulus presentation the experimenter - giving the subject all the time he wants - may choose to increase the number of alternative stimuli among which the subject has to discriminate and determine the critical set N_c, beyond which errors begin to occur. Under these circumstances the $\log N_c$ bits define the channel capacity.

A tremendous number of experiments using the above methodology has been performed in the past 40 years and a very lucid summing up of some most representative cases has been offered by George A. Miller (1974) in a, by now, classic collection of seven essays. In essence these essays refer to three distinct attributes of a human subject - seen as an information processor: the determination of the span of immediate attention or short-term memory; the span of absolute judgement; and the limit of self-embedding in a natural language.

Let us quickly review Miller's results:

A. The span of immediate attention

This attribute refers to the maximum number of digits (the longest string) a normal adult can repeat without error. This number is about seven to eight digits. This "memory span" is limited by the number of symbols and not by the amount of information that these symbols may represent.

B. The span of absolute judgement

This has to do with the maximum number of alternative stimuli beyond which confusion and errors are reported. The alternative stimuli may differ mutually in one or more parameters (dimensions). No matter how many alternative stimuli (differing along one dimension only - say pitch) the subject is asked to evaluate, the best we expect him to do is to assign them to about 6-7 different classes without error. Or, if we know that there are N alternative stimuli, then the subject's judgement enables us to narrow down (or "compress") the particular stimulus to one out of $\sim N/7$.

The addition of dimensions to the stimulus (say pitch and loudness) increases the channel capacity but at a decreasing rate (this happens even when the dimensions along which the stimuli differ are not independent). In other words, we increase the total capacity at the expense of accuracy for any particular dimension (parameter); so we can make relatively crude judgements of several things simultaneously.

It is tempting to assume that in the course of evolution the success of the organism has to do with the ability to respond to the widest possible range of stimuli. In order to survive in a fluctuating environment it is better to have little information about a lot of things than to have a lot of information about few things (e.g. a restricted part of the environment).

At any rate it seems that there is a clear and definite limit in the accuracy with which we can absolutely identify the magnitude of a uni-dimensional stimulus variable: we call this limit, according to Miller, the "span of absolute judgement". This is usually somewhere in the neighbourhood of seven alternative stimuli (or $\log_2 7 \sim 2.5$ bits).

So there is a span of absolute judgement that can distinguish about seven categories and there is a span of attention that can encompass about seven objects or symbols at a glance. Miller is quick to appreciate however that in spite of the appearance of the "magical number seven plus-minus two" in both human attributes, the span of absolute judgement and the span of immediate memory are quite different kinds of limitations that are imposed on our ability to process information: absolute judgement is limited by the amount of information while immediate memory is limited by the number of items.

Using hierarchical coding we can, on principle, squeeze more and more bits on a given item. So, the span of immediate memory seems to be almost independent of the number of bits per symbol; with proper coding our ability to process information in terms of channel capacity may very well increase from 2.5 to 10 or 25 bits (Miller (1974)).

In the words of Miller the "magical number seven" may very well be only "a pernicious Pythagorean coincidence".

In spite of the above cautionary remarks we intend - referring to a given hierarchical level of cognition - to pursue the above "coincidence" as a symptom of a more deep and fundamental underlying dynamics.

The fact is that our pitifully small capacity for processing information has made it necessary for us to develop ways of "recoding" or abstracting the essential features of the perceived stimuli in a hierarchical fashion, thereby expressing our experiences in the form of simple "laws" or with brief words: it is this demonstrable ability to dynamically store and subsequently "compress" variety or to map information from higher to lower dimensionality state space, which gives us the impetus to propose (next section) a dynamical model which on principle takes care, in one stroke, of several seemingly unrelated attributes of human information processing.

C. The recurrent (self-referential) character of natural (and artifical) languages, and the maximum number of discriminable self-embeddings

As almost everybody knows, clichés aside, learning a language does not involve memorization of particular strings or symbols but rather knowledge of rules of generating admissible strings of symbols.

These rules are given in terms of highly non-linear recursive algorithms. In artifical languages we do possess also recursive subroutines, whose use aims at increasing resolution via successive iterations.

Occasionally the main routine is interrupted to perform a subroutine and then we return to the main routine when the subroutine is finished. It is impossible also for one subroutine to refer to another subroutine. Let us see, however, what happens when a subroutine refers to itself. The "computer" interrupts the main routine to go into subroutine S. In the middle of the execution S, however, before S is completed, the "computer" is instructed to stop what it is doing and execute subroutine S. No difficulty arises until subroutine S is completed and the "computer" must decide where to resume its work. Unless special attention is taken in writing the programme, the "computer" will not be able to remember that the first time it finishes subroutine S it must re-enter subroutine S and the second time it finishes subroutine S it must return to its main routine. The situation gets more difficult with subroutine S calling on itself repeatedly; for, each time it does so, a new re-entry address has to be recalled.

It is also a feature of natural language that sentences can be inserted inside sentences. Miller gives, amongst others, the following example:

"The King who said "My Kingdom for a horse" is dead" (one embedding).

"The person who cited "The King who said "My Kingdom for a horse" is dead" as an example is a psychologist" (two embeddings).

"The audience who just heard "The person who cited "The King who said "My Kingdom for a horse" is dead" as an example is a psychologist" is very patient" (three embeddings).

As the recursive process goes on beyond, say, the fourth or the fifth self-embedding, "chaos" sets in literally and actually; human subjects are thus very poor also in dealing intelligibly with more than a few recursive interruptions.

We may ask now: is there any common feature in the "linguistic" attributes A, B and C? In this section we intend to propose a dynamical model of an information processing scheme which ensures compressibility of (environmental) strings of digits and also fixes an optimum resolution (that is an optimum code word-length) which maximizes storage capacity. In specific examples we are going to see that the optimum (average) length of such a code is of the order of four to eight bits. Furthermore, the proposed scheme ensures that increasing resolution beyond a certain limit amplifies just microscopic noise, thereby destroying compressibility.

9.2. Information processing (generation-dissipation) and memory of a chaotic processor

Reliable information processing rests upon the existence of a "good" code or language: namely, a set of recursive rules which generate and store variety (strings of symbols) at a given hierachical level and subsequently compress it, thereby revealing information at a higher level.

To accomplish this task, a language - like good music - should strike an optimum ratio of variety (stochasticity) versus the ability to detect and correct errors (memory) at every moment. Is there any dynamics which might model this dual objective in state space available today. The answer is: On principle, Yes (see for example Nicolis (1982a, b, 1984, 1985), Nicolis et al. (1983)). Consider the class of dynamical systems described by at least three coupled first-order ordinary non-linear differential equations whose repertoire includes (for different sets of values of the control parameters) multiple steady states, stable periodic orbits (limit cycles), tori and strange chaotic attractors. We may consider that variety to be generated when the volume in the state space of the system expands along certain directions through the dynamical evolution (thereby decreasing resolution) and to be compressed when the volume in state space occupied by the flow contracts (increases resolution) along other directions towards a "compact" ergodic set - the attractor. A strange attractor of a three-dimensional system, for example, creates variety along the direction of its positive Lyapunov exponent λ_+ and constrains variety (thereby revealing information) along the direction of its negative Lyapunov exponent λ_-. Since $|\lambda_-| > |\lambda_+|$, the attractor is acting literally as a "compressor", <u>on the average</u>[1)]

The instability of motion on a chaotic attractor is conveniently illustrated in the one-dimensional map on the interval which is constructed as a Poincaré return map of the attractor. This type of "analogue-digital conversion" is accomplished by parametrizing the attractor along a one-dimensional cut and plotting the position a trajectory crosses the cut versus the position it crosses the next time around the attractor. This may give rise to a Markov chain whose number of states depends on the partition of the interval.

Following Shaw (1981), we point out that the change in observable information is generally given by the $\log_2$ of the ratio of states $\Sigma(t)$ indistinguishable before and after some time interval

$$\Delta I = \log_2 \frac{\Sigma_f}{\Sigma_i} \tag{99}$$

or

$$\sim \log_2 \frac{V_f}{V_i} , \tag{100}$$

where V_f, V_i are the final and initial volumes in state space. The rate of in-

1) The Lyapunov exponents determine the <u>average</u> amount of information production ($\lambda > 0$) or dissipation ($\lambda < 0$) per time unit.

formation creation or dissipation is given as

$$\frac{dI}{dt} = \frac{1}{V}\frac{dV}{dt} \sim \frac{1}{\Sigma}\frac{d\Sigma}{dt} \quad . \tag{101}$$

In non-chaotic systems the sensitivity of the flow on the initial conditions grows with time at most polynomially, say as

$$\Sigma(t) \sim t^n \quad ; \tag{102}$$

then

$$\frac{dI}{dt} \cong \frac{n}{t} \quad . \tag{103}$$

The variety creation rate from such a system converges to zero as time passes, so the behaviour of the system is predictable. Concomitantly, the set of generated strings is formed by rational numbers.

In chaotic systems the sensitivity of the flow on the initial conditions grows exponentially with time, say as

$$\Sigma(t) \sim e^{nt} \tag{104}$$

and

$$\frac{dI}{dt} \sim n \quad . \tag{105}$$

Such a system is a continuous information source. This information is not implicit in the whatever initial conditions but is generated by the flow itself in state space: a chaotic dynamics creates strings of digits which constitute aperiodic oscillations or segments of irrational, non-computable numbers. If such a chaotic system is observed by another cognitive system at a sampling rate less than n, no prediction of the system is possible. We are interested here in cognitive systems characterized by non-conservative compact flows, where the phenomenon of attraction, that is of dimensionality compression (see next section), is possible. Let us argue in terms of an appropriately constructed Poincaré map for the flow. For a typical one-dimensional map in the interval $y = F(x)$ the probability density $P(x)$ for finding an orbit at x can be estimated via successive iterations from the recursive relation

$$P_n(x) = \sum_{i=1}^{2^{n-1}} \frac{P_1(x_i)}{\left|\frac{dF^{(n-1)}}{dx}\right|_{x_i}} \quad , \quad x_i = (F^{n-1})^{-1}(y) \quad , \tag{106}$$

where the first p.d.f. $P_1(x)$ can be taken arbitrarily and the resulting $P(y)$ - corresponding to the initial $P_1(x)$ transformed by the action of the map -

- is the first approximation. Successive iterations of the map will produce closer and closer approximations to the correct asymptotic function of P(x). If the map possesses a stable steady state or a number of stable periodic orbits, P(x) will converge to a sharp δ-function spike or a series of δ-like functions on the periodic points. In the chaotic regime of the control parameters the map will possess a continuous P(x) with interspersed spikes.

The information change ΔI per iterations of the particular initial point x on the interval will be determined as

$$I = \log_2 \left| \frac{dF}{dx} \right| \tag{107}$$

and will amount to variety creation for slopes > 1 and information (dissipation) for slopes < 1.

The <u>average</u> information change over the whole interval $0 \leqslant x \leqslant 1$ will be given by

$$\langle I \rangle = \int_0^1 P(x) \log_2 \left| \frac{dF(x)}{dx} \right| dx \quad \text{bits} . \tag{108}$$

If t(x) stands for the time between successive crosses through the return map, the average information production (or dissipation) rate can be defined as

$$\left\langle \frac{dI}{dt} \right\rangle = \int_0^1 \frac{P(x)}{t(x)} \log \left| \frac{dF(x)}{dx} \right| dx \quad \text{bits/sec.} \tag{1o9}$$

Finally, if the initial condition from which the process starts is not known exactly (which is always the case[1)]), i.e. it is determined by an a priori probability density distribution $P_0(x)$, one gets, for the information value of this initial condition, the expression

$$S = \int_0^1 P_0(x) \log_2 \left[\frac{P_0(x)}{P(x)} \right] dx \quad \text{bits} . \tag{110}$$

Then the "memory" of the processor measured as the time elapsing until the flow gets completely disconnected from the initial conditions can be defined as

1) This initial condition is a point x_0 on the unit interval. It can be expressed as a digital number .0110011100111001...; with overwhelming probability this number is incompressible.

$$T = \frac{S}{\langle \frac{dI}{dt} \rangle} = \frac{\int_0^1 P_0(x) \log_2 \left[\frac{P_0(x)}{P(x)} \right] dx}{\int_0^1 \frac{P(x)}{t(x)} \log_2 \left[\frac{dF(x)}{dx} \right] dx} \quad \text{sec.} \tag{111}$$

Beyond this limit, the iterative process becomes so fine that the processor amplifies intrinsic microscopic noise essentially. Due to the great resolution created by the cascade of iterations, this noise cannot be "smeared out" any longer and declares itself on the cognitive level, thereby rendering any further measurement or observation unreliable.

Computing simulation (Shaw (1981)) shows that in specific cases the number of iterations by which T is achieved may be rather small ($\sim$ 20 ÷ 30).

The information dimension D_1 of a strange attractor (see, for example, Farmer (1982), Farmer et al. (1982)) is the number of bits one requires for the unambiguous determination of any point on the attractor. Alternatively, it represents the maximum number of bits we can reliably store in such a self-sustained dynamics under a given resolution.

The "average degree of compressibility" ensured by the attractor is defined (Nicolis et al. (1983)) as

$$\eta = \frac{N - D_I}{N} \; \% \quad , \tag{112}$$

where N is the dimensionality of the state space in which the attractor is embedded as a compact subset.

Formally, in order to calculate D_I one proceeds as follows: one partitions the N-dimensional state space into $n(\varepsilon)$ hypercubes of size (degree of resolution) ε, from each of which the flow (or "the trace of the scanning path" of the processor) passes at least once. Let $P_i(\varepsilon)$ be a number proportional to the relative frequency with which the scanner visits the specific square i. Then the "entropy" of the digitized pattern can be calculated as

$$- \sum_{i=1}^{n(\varepsilon)} P_i(\varepsilon) \log_2 P_i(\varepsilon) \quad \text{bits} \quad , \tag{113}$$

and as the resolution becomes finer and finer one may be interested in calculating how the distribution $P_i(\varepsilon)$ scales with resolution. One could obtain then the "information dimension" or the bits required for the determination of any point on the "scanning curve" of the processor as

$$D_I = \lim_{\varepsilon \to 0} \frac{-\sum_{i=0}^{n(\varepsilon)} P_i(\varepsilon) \log_2 P_i(\varepsilon)}{\left| \log_2 \frac{1}{\varepsilon} \right|} \qquad (114)$$

i.e. the asymptotic value of the slope of entropy versus resolution.

Actually, it turns out that the calculation of D_I is better feasible if, instead of starting from the above expression, one uses the relationship between the information dimensionality and the spectrum of Lyapunov exponents of the flow (or the discrete map). In addition, it is important to calculate not only D_I but also the standard deviation ΔD_I as the set of second moments of a distribution whose first moments are the Lyapunov exponents. Such a quantity would indicate how much fuzziness enters in the average degree of compressibility of $\eta = (N - D_I)/N$ of information, realized by the attractor in N-dimensional space.

For an attractor simulating some aspects of a cognitive system one has two basic requirements: big storage capacity and good compressibility. Let us turn first to the simplest attractors (in any N-dimensional state space), namely stable steady states and stable limit cycles: they have information dimensionality zero and one, respectively. Consequently, they are very poor as information storage units, but since they possess only non-positive Lyapunov exponents, ideal as information compressing devices.

Strange attractors on the other hand, via a harmonious combination of positive and negative Lyapunov exponents λ_+, λ_-, can on principle comply somewhat with both requirements: they may possess a considerable information dimension and this idiom makes them suitable for information storage. On the other hand, being "attractors", i.e. possessing also negative Lyapunov exponents λ_- such that $|\lambda_-| > \lambda_+$, they can serve at the same time as information compressors.

The type of the strange attractor we intend to employ as a model here should be rather "loose" in order to comply with the preceding experimental results. It must, in other words, possess a large information dimensionality in order to allow for maximum dynamical storage of variety.

We are going to investigate the optimum resolution ε under which this information dimensionality becomes maximum. The ensuing compressibility factor $C_I = N - D_I$ will also be investigated. Summarizing, in case our primary aim is compressibility, the best attractors around are steady states and limit cycles. In case, however, our data ask for a model which can ensure first of all large storage ability without jeopardizing compressibility, a strange attractor is the preferred model. In cognition one presumably needs both types of attractors working in tandem. Nicolis and Tsuda (1985) have shown that we can use just one attractor which, under the action of external noise, switches from one model to the other, thereby complying with both requirements.

9.3. Specific examples and relevance of the model

What is the relevance (if any) of the dynamical theory given in Section 9.2. as against the experimental results reported in Section 9.1, which essentially refer to empirical limitations of human attributes in information processing? To answer this question let us try to see if there exists, in various specific cases of dynamical models of processors, an optimum resolution ε (different for each model) or an optimum codeword length for which the cardinal parameter in our theory, the information dimension of the attractor, becomes maximum - without jeopardizing the average compressibility factor $\eta = (N - D_I)/N$.

Let us consider the expression $C_I = N - D_I$ representing the degree of average compression realized by the particular attractor and try to calculate a critical resolution ε^* for which D_I becomes maximum or

$$\frac{\partial C_I}{\partial \varepsilon} = 0 \quad . \tag{115}$$

To take into account the best case let us perform the above calculation for the maximum value of

$$D_I = \frac{\sum_{i=1}^{n(\varepsilon)} P_i(\varepsilon) \log_2 P_i(\varepsilon)}{\log \frac{1}{\varepsilon}} \quad , \quad \varepsilon \text{ finite} . \tag{116}$$

The maximum value of the information dimension under the given finite resolution ε is the "fractal dimension" of the attractor and it is realized for

$$P_i(\varepsilon) = \text{const} = \frac{1}{n(\varepsilon)} \quad . \tag{117}$$

So,

$$D_I = -\frac{\ln[n(\varepsilon)]}{\ln \varepsilon} \tag{118}$$

and

$$C_I = N + \frac{\ln[n(\varepsilon)]}{\ln \varepsilon} = \frac{N \ln \varepsilon + \ln[n(\varepsilon)]}{\ln \varepsilon} = \frac{\ln[\varepsilon^N n(\varepsilon)]}{\ln \varepsilon} \quad . \tag{119}$$

Now we intend to express $n(\varepsilon)$ in terms of the Lyapunov exponents of the attractor and the resolution length ε. To this end, we proceed as follows. Let the number of points on a (one-dimensional) Poincaré return map of the attractor concerned be M; this means that we determine the orbit with strings of M digits long. The number of cells representing the attractor, with a degree of coarseness ε, is just $n(\varepsilon)$. Now, for the degree of resolution ε to be meaningful, it is necessary that each of the possible outcomes of an orbit M digits long could be determined with precision. Since we deal here with systems whose dynamics can be adequately represented in each step by two possible states, say 0 and 1, the number of outcomes is clearly 2^M. Hence ε should be chosen such that

$$\varepsilon \sim \frac{1}{2^M} \quad . \tag{120}$$

On the other hand, let t_c be the sampling time or, alternatively, with sampling time period $\Delta t = 1$, the number of sampling (iterations) probing our attractor. For an attractor represented by M points, it is therefore meaningful to choose

$$\frac{n(\varepsilon)}{M} \sim t_c \quad . \tag{121}$$

Owing to the chaotic character of the dynamics, after t_c iterations in the map the system is completely disengaged from the initial condition. This means that t_c can be estimated from the relation

$$\varepsilon e^{\lambda_+ t_c} \sim 1 \tag{122}$$

where λ_+ is the positive Lyapunov exponent. So

$$t_c \sim \frac{\ln \frac{1}{\varepsilon}}{\lambda_+} \quad , \tag{123}$$

and finally

$$n(\varepsilon) = \frac{\left(\ln \frac{1}{\varepsilon}\right)^2}{\lambda_+ \ln 2} \quad . \tag{124}$$

Substituting in the expression of C_I we get

$$C_I = N + \frac{2\ln(\ln \frac{1}{\varepsilon}) + \ln\left(\frac{1}{\lambda_+ \ln 2}\right)}{\ln \varepsilon} \tag{125}$$

or

$$C_I = N + \frac{\gamma + 2\ln\left(\ln \frac{1}{\varepsilon}\right)}{\ln \varepsilon} \tag{126}$$

where

$$\gamma = \ln\left(\frac{1}{\lambda_+ \ln 2}\right) .$$

The reqirement for maximum fractal dimension $\partial C_I / \partial \varepsilon = 0$ gives for the optimum resolution

$$\varepsilon^* = \exp\left[-e^{(2-\gamma)/2}\right] \tag{127}$$

or the optimum code length

$$M^* = \frac{\ln \frac{1}{\varepsilon^*}}{\ln 2} , \tag{128}$$

from which

$$C_I(\varepsilon^*) = N - 2e^{-(2-\gamma)/2} \text{ bits} . \tag{129}$$

These are our final formulas for the minimum average compressibility, i.e. the maximum fractal dimensionality of the attractor concerned and the corresponding optimum code length M^*.

It remains to make sure that the extremum of the function

$$f(\varepsilon) = -D_I(\varepsilon) = \frac{\gamma + 2\ln\left(\ln \frac{1}{\varepsilon}\right)}{\ln \varepsilon} , \qquad \varepsilon \neq 0, \tag{130}$$

is indeed a minimum or $D_I(\varepsilon)$ is maximum. In fact we have

$$\frac{\partial f(\varepsilon)}{\partial \varepsilon} = \frac{2 - \left[\gamma + 2\ln\left(\ln \frac{1}{\varepsilon}\right)\right]}{\varepsilon (\ln \varepsilon)^2} \tag{131}$$

and

$$\frac{\partial^2 f(\varepsilon)}{\partial \varepsilon^2} = \frac{-6 + 2\gamma + 4\ln\left(\ln \frac{1}{\varepsilon}\right) + (\ln \varepsilon)\left[\gamma - 2 + 2\ln(\ln \frac{1}{\varepsilon})\right]}{\varepsilon^2 (\ln \varepsilon)^3} ; \tag{132}$$

for $\varepsilon = \varepsilon^* = \exp\left[-\exp\left(\frac{2-\gamma}{2}\right)\right]$ we get

$$\left(\frac{\partial^2 f(\varepsilon)}{\partial \varepsilon^2}\right)_{\varepsilon = \varepsilon^*} = \frac{2}{\exp\left[-2\exp\left(\frac{2-\gamma}{2}\right)\right] \exp \frac{3}{2} (2-\gamma)} \tag{133}$$

which is always positive. So $f(\varepsilon)$ possesses a minimum at ε^* and consequently $D_I(\varepsilon) = -f(\varepsilon)$ becomes maximum. $D_I(\varepsilon)$ becomes zero for $\varepsilon_0 = \exp\left[-e^{-\gamma/2}\right]$. Negative values of D_I do not possess any physical meaning, of course.

Notice that the value of $D_I(\varepsilon^*)$ is much less than the fractal dimensionality D_I calculated in the limit $\varepsilon \to 0$. As a matter of fact, $D_I(\varepsilon^*)$ does no longer refer to a geometrical object. Rather, it provides us with a convenient measure of the efficiency of an information-producing device.

Let us now apply formulas (128) and (129) to a number of simple cases:

A. The Lorenz attractor $\dot{x} = \sigma(y - x)$, $\dot{y} = y + rx - xz$, $\dot{z} = xy - bz$ (for $b = 4$, $\sigma = 16$ and $r \simeq 45.92$)

With the computer we calculate $\lambda_+ \sim 1.5$, $y \sim -0.03895$, $C_I(\varepsilon^*) \sim 3 - 0.72157 \sim 2.28$, $\varepsilon^* \sim 0.062$ and $M^* \sim 4$. That is, the optimum code maximizing storing capacity by the Lorenz attractor is ~ 4 bits long. Nevertheless, the resulting compressibility is impressive too - much larger than what one calculates from the information dimensionality with $\varepsilon \to 0$.

$[C(0) = 3 - 2.06 = 0.94.]$

B. The Bernoulli map: $x' = \mathrm{Mod}_1(2x)$

In this case of a one-dimensional map - assumed here to be uniform - the number of covering segments of resolution ε is

$$n(\varepsilon) \sim \frac{\ln(\frac{1}{\varepsilon})}{\lambda_+} , \tag{134}$$

from which we obtain

$$C_I \sim \frac{N + [\ln(\ln \frac{1}{\varepsilon}) - \ln \lambda_+]}{\ln \varepsilon} . \tag{135}$$

The requirement $\partial C_I / \partial \varepsilon = 0$ gives for the optimum resolution

$$\varepsilon^* \sim \exp \left[-e^{1+\ln \lambda_+} \right] , \tag{136}$$

from which

$$C_I(\varepsilon^*) = N - e^{-(1+\ln \lambda_+)} . \tag{137}$$

The numerical calculations for $\lambda_+ = 2$ in this case give $C_I(\varepsilon^*) = 0.81606$, $\varepsilon^* = 0.004354$ and $M^* \sim 7$ to 8 bits.

Summarizing, the strange attractor models show that the trajectories become acausal after relatively few iterations on a map, i.e. they get completely disconnected from the initial conditions. This allows internally generated or environmental random noise to amplify and declare itself, so to speak, on mascoscopic levels. So, qualitatively we justify the observed fact that beyond a limited number of self-embeddings the strings or "words" in a certain natural or artificial language appear really random.

9.4. Outlook

Let us end up with some remarks concerning the basic issue of the number of categories from which a given cognitive system can make unambiguous classification of "code-words". In English, for example, the average word has about 5 letters. In view of the redundancy of English language ($\sim$ 75%) we do know that the average information per letter is $\sim$ 2 bits. So the information content of the average word in the English language is about 10 bits.

Now, what is this supposed to mean? It means that if one tries to guess an English word fixed by the experimenter one should be able to do it in 10 guesses involving "yes-no" answers.[1)] In turn, this means that guessing has to do with narrowing down logical categories. Now, here is the rub: one has to select or sort out from a well-defined set. Suppose that instead of the above example you are asked to sort out a particular arrangement of a deck of cards (Rappoport (1956)).

Since each arrangement is equiprobable, its probability being $\sim 1/52!$, it follows that $\log_2(52!)$ "yes-no" decisions are required to reconstruct a particular order from a shuffled deck. Now $\log_2(52!) \sim \log_2 10^{68} \sim 200$ bits, so one would think that the specific order or string can be deduced in about 200 "yes-no" guesses. The real issue, however, is that in this case we do not possess the linguistic tool to classify the card arrangements so as to be able to make judicious dichotomies for "yes-no" answers. Our knowledge of card arrangements is confined to a very small subset of "compressible" sequences, that is, of arrangements which are reproducible with fewer instructions than their own length. To reconstruct any other arrangement amongst the incompressible ones one has to <u>name</u> the position of each card or symbol separately. It turns out then that the difficulty of <u>classifying</u> springs from the difficulty in <u>compressing</u> or devising algorithms which can reproduce a given string of symbols with fewer bits. <u>A cognitive gadget ensuring maximum compressibility without jeopardizing storage capacity provides then the optimum classification scheme, that is the optimum size of the set.</u>

---

1) It is of course futile to try this methodology in the case where the cognitive gadget does <u>not</u> possess an attractor for this particular word.

10. Intermittency in perception and communication processes

Taking averages in physical sciences in general and in communication theory in particular results always in some selective loss of detail. If it happens that a few details account practically for the whole pattern, then the averaging process simply "washes out" all the essential information. In statistical mechanics, on the one-hand, the pursuit of evolution of the microscopic probability density function and its moments through the formalism of the Master equation and Fokker-Planck equation in systems far from equilibrium and near bifurcation points manifests the "break down" of the law of large numbers; this has been amply demonstrated in recent years (Haken (1978), Nicolis and Prigogine (1977)) - together with the ensuing invalidation of the "mean field regime". The entropy for example is just the mean value of the distribution $-\ln p(x)$ where $p(x)$ stands for the a priori probability density distribution of a (finite or infinite) set of elements constituting a certain pattern. Some of the elements of the set may be extremely improbable w.r.t. a certain observer or prone to deliver a disproportionally large amount of information upon reception. Of particular interest is the case where the median value of $p(x)$ is the least probable one. In such cases the usual expression(s) for the static entropy

$$S = -\sum_i p_i(x) \log_2 p_i(x) \tag{138}$$

or

$$\Delta S = -\int p(x) \log_2 p(x) dx \quad \text{(in bits)} \tag{139}$$

is perhaps inadequate in characterizing quantitatively the information transaction.

In this section we intend to treat dynamical systems where the variety production or information dissipation are given by the dynamical analogues of the entropy and are couched in terms of the Lyapunov exponents of the flow or the discrete map concerned.

In the following we briefly review some experimental evidence about the dynamics of visual pattern perception and recognition (which are the "crucial features" of the pattern in such a case and how is the processor dealing with them?) as well as the irrelevance of the "law of averages" in certain "coin tossing" and communication problems. We provide some evidence that information processing of patterns can be associated with a set of selective "jumps" from complex feature to complex feature (i.e. a feature with great curvature). Around such a feature the receptor displays the greatest sensitivity, i.e. the largest differential output (in terms, say, of firings per time unit) for any given (small) differential input - and an accompanying instability rendering the scanning process chaotic.

So, perception (and perhaps cognition at another, higher level) is not so much involved with averaging as with selection of singularities in the pattern under investigation. These singularities hold the "Lion's share"; the rest of the features are simply irrelevant in the sense that small "saccadic" eye deviations around them are provoking commensurately small difference in response. So irrelevant features are causing smooth responses while relevant ones are provoking unpredictable (chaotic) responses.

10.1. The break down of the "law of large numbers" in pattern recognition and random walks

I. Review of some experimental results in visual pattern perception (Noton and Stark (1971a, b), Noton (1970))

Which are the "crucial features" of a visual pattern and how is the optical (human) cortex (in cooperation with the sensor (fovea[1]) and the optical muscular apparaturs) dealing with them? In a remarkable series of experiments conducted in the early seventies, Noton and his associates performed a number of investigations which are discussed below. To begin with, just look at Fig. 14. The left part represents the famous bust of Queen Nefertiti. On the right, the trajectory of the eye movement of a human observer is displayed (as recorded by

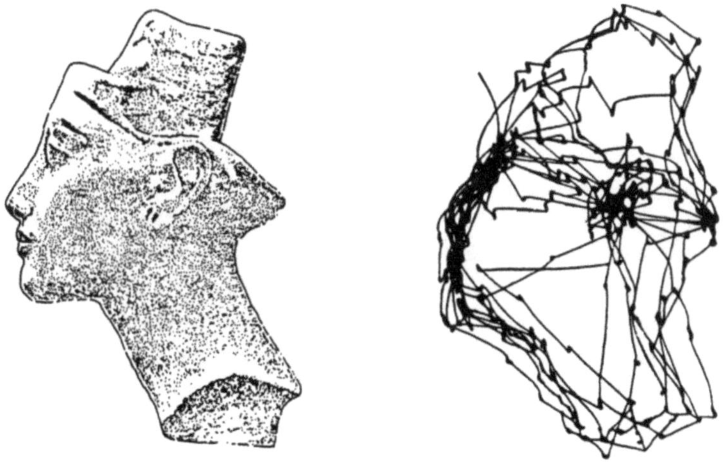

Fig. 14 Scanning Nefertiti's bust (after Noton and Stark (1971))

---

[1] The fovea constitutes a small central area of the retina and is characterized by the highest concentration of photoreceptors.

A.L. Yarbus of the Institute for problems of Information Transmission in Moscow, USSR) - as it was scanning Nefertiti's head and neck in the process of perception. The experiments were performed in an attempt to settle the controversy of Gestalt (parallel, global) versus sequential (step by step - or iterative) pattern recognition. (An essential prerequisite for the meaningfulness of the attempted comparison is, of course, that the pattern must be extended enough in space in order to allow for scanning by the sensor.). The result of the experiments heavily supports the hypothesis of serial, piecemeal perception and recognition. Specifically two questions are answered: a) which are the features of the pattern that the optical cortex selects as the key items for identifying the object? and b) how are such features integrated and related to one another to form the complete internal representation of the object?

First of all we have good evidence that the optical system is a hierarchical one (Noton (1970)), and that mappings take place between the successive hierarchical levels, thus giving rise to feedforward-feedback loops. The higher levels (cortex) receive information from the lower levels and respond by sending commands to the muscular apparatus of the sensor either to move in order to "phaselock" with the feature under investigation or to move away and assume the next algorithmic step of scanning the pattern. The experiments demonstrate that the above scanning algorithm is far from being smooth and "homogeneous". There are parts on the pattern that hold for the processor most of the information about the subject. The fixation or the "holding time" of the receptor tend to cluster around the parts characterized by sharp curves (small radius of curvature), angles, or, in general, around areas where the curve is "non-differentiable". It appears that the angles are the principal features the brain uses to store and recognize the pattern. (There is also neurophysiological justification for such a preference revealed through the painstaking experiments of Hubel and Wiesel; it appears that there are angle-detecting neurons on the frog's retina and angle-detecting neurons in the visual cortex of cats and monkeys. Recordings obtained from the human visual cortex by Marg (Noton and Stark (1971a)) of the University of California at Berkeley give indications that this result can be extended to the human visual cortex as well.) The scanning of Fig. 14 bears witness to the heavy preference of the processor to areas of the face characterized by sharp curves. Such features are <u>complex</u> (they need "many bits" for their most laconic description and are therefore endowed with high information content). Dynamical analysis of the sequence of fixation of the receptor at the different states-"features" of the pattern suggests a format for the interconnection of these "states" into the overall internal representation. The result of this experiment reveals that the sensor directed by the brain is essentially involved with two types of "scanning pathways": Regular and irregular. Specifically it appears, as Noton and Stark (1971a) put it, that "(the) eyes usually scanned it (the pattern) following - intermittently but repeatedly - a fixed path which we have termed their 'scan path' ... . The occurrences of the scan path were separated by periods in which the fixations were ordered in a less regular manner ... Scan paths usually occupied from 25 to 35 percent of the subject's viewing time, the rest was devoted to less regular eye movement". Following Noton and Stark we may conclude then that the internal representation (or the mapping) of a pattern in the memory system is an assemblage of features (states) mediated essentially by a feedback loop: A sequence of senso-motor traces recording, abstracting and subsequently rèporting to the brain a "Markov chain" as it were, of states-features - whereupon the brain directs the next move of the peripheral activity. The time intervals during which the system holds a given state have obviously to do with the excitation of the directing neuronal tissue - which in turn depends on the degree of curvature of the feature-state involved (Noton (1971)). This Markov chain, this algorithm, is subsequently stored in the brain isomorphically as, say, a pattern of circulating electrical activity. When the observer is subsequently encountering the same pattern again he recognizes it by matching it with the "feature-ring" or the Markov chain which constitutes the internal representation in his memory, state-by-state. Matching then or recognizing consists in calculating the "distance" or the cross-correlation between two Markov chains: The memorized one and the one which runs as the observer re-examines the pattern. Obviously, the first chain directs the steps of the second through the brain-sensomotor loop and so learning takes place: Beyond a few re-runs the object becomes "familiar" as the crosscorrelation above tends to one. It would be then, under such circumstances, rather improper to try to gather the information conveyed from observing Nefertiti's bust by calculating the "entropy" of this pattern. By the way, this could be done as follows: You partition the two-dimensional pattern in Fig. 14 into $n(\varepsilon)$ squares of size ε through each of which the trace of the scanning path passes at least once. Let p_i be a number proportional to the relative frequency with which the scanner visits the specific square i. Then the "entropy" of the pattern could be calculated as

$$- \sum_{i=1}^{n(\varepsilon)} p_i \log_2 p_i \quad \text{bits} \tag{140}$$

and as the resolution ε becomes finer and finer one may be interested in calculating how the distribution $p_i(\varepsilon)$ scales with resolution. In that case one could obtain the "information dimension", or the bits required for the determination of a point of the scanning curve, as

$$D_I = \lim_{\varepsilon \to 0} \frac{-\sum_{i=1}^{n(\varepsilon)} p_i \log_2 p_i}{\left| \log_2 \frac{1}{\varepsilon} \right|} , \quad (141)$$

i.e. the asymptotic value of the slope of entropy versus resolution ($\| \log \varepsilon \|$).

Of course, it is implicitly assumed that Nefertiti's bust constitutes indeed a "strangely attracting object" or, that the scanning path remains bounded in space and continuously iterating, on the basis of a dynamical control algorithm dictated by the brain centers - which "fire" more or less frequently depending on the curvature of the encountered last detail (state). In view of the shape of the scanning trace on the right side of Fig.14, instead of relying on (141), one could obtain a more meaningful collective property or quantitative measure of the information gathered by investigating the standard deviation of D_I. However, to make calculations feasible we should not start from formula (141) but rather from a relationship between the information dimensionality and the spectrum of Lyapunov exponents of the flow or the discrete map, and calculate the standard deviation of D_I, ΔD_I, from the NUFs of the corresponding Lyapunov exponents. The NUF of D_I would indicate how much fuzziness enters in the degree of compressibility $(N - D_I)/N$ % of information realized by the attractor in the N-dimensional space[1].

For example, in the case of the immune system - envisaged as an information processor - a tremendous variety of templates (antibodies) is put forth, either spontaneously or in response to stimulation by a single antigen; the "polyclonicity" of these templates may provoke an autoimmune effect; as a result the organism may suffer more damage in terms of entropy production from such an excessive response of the processor than from the initial intruder. We may say in such a case that the NUF of D_I by far exceeds the compressibility factor $N - D_I$. In case of "monoclonal" antibodies, on the other hand, one should expect the NUF of D_I to be just zero.

II. Information transfer through intermittent channels or "error clustering" media

Let us now change gears and refer to another seemingly unrelated example concerning again the break down of the "law of averages" in Markov chain sequences. The example has essentially to do with cointossing games or random walks and has been elevated into the prominence of a new scientific paradigm by Feller (1970) and subsequently by Berger and Mandelbrot (1963). According to "widespread beliefs" (the characterization is due to Feller) a so-called "law of averages" should ensure that, in a long cointossing game, each player will be in the winning side for about half the time and that "switching" will take place fre-

---

1) In the present work we did calculate the NUF for the dominant λ of maps and flows (Section 4).

quently from one player to the other. Fig. 15, taken from Feller, represents the result of a computer experiment simulating 10,000 tosses of a fair coin. The top line contains the graph of the first 550 trials; the next two lines represent the entire record. The surprising thing has to do of course with the length of intervals between successive crossings of the "zero"-axis. On an isomorphic basis you can imagine these graphs as standing for a one-dimensional random walk without barriers and ask how often the walker is likely to change sides. Because of the symmetry of the walk one expects that in a long walk a man should spend about half of his time on each side of the starting spot. Exactly the opposite is true. Regardless of how long he walks, the most probable number of changes frome one side to the other is 0, the next probable 1, followed by 2, 3 and so on. If a man walks one step every second say for a year, in one out of 20 repetitions of this experiment the walker could expect to go along one direction for more than 364 days and 1 h. (In such simulations one is not using coins of course but decides on the basis of the upcoming digits of, say, 100 decimals: An even digit is "0" and an odd is "1".) In conclusion it is quite likely that, in a long coin-tossing game, one of the players remains on the winning side practically the whole time, the other on the losing side. (In the experiments of Fig. 15 mentioned above, in 10,000 tosses of a perfect coin the lead is at one side for more than 9930 trials and at the other for fewer than 70 with probability greater than 10%.)

On the other hand, one "should expect" that in a prolonged coin-tossing game the observed number of changes of lead should increase roughly in proportion to the duration of the game. This again is false.

Feller proves that the number of changes of lead in n trials increases only as $\sqrt{n}$; so, in 100 n trials one should expect only 10 n times as many changes of lead as in n trials. Putting it in another way, if N_n is the num-

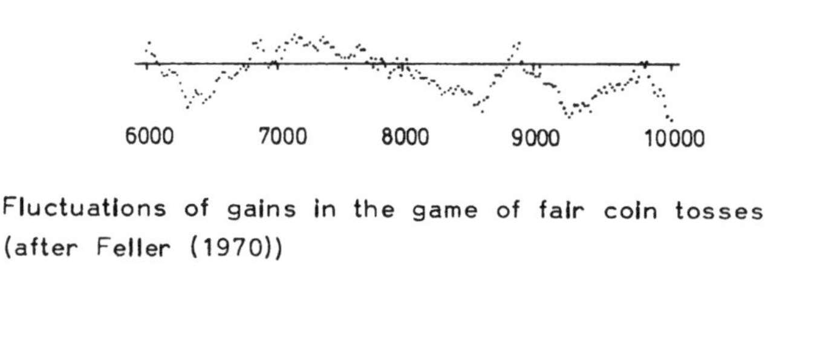

Fig. 15 Fluctuations of gains in the game of fair coin tosses (after Feller (1970))

ber of changes in lead,

$$\frac{N_n}{n} = \frac{\sqrt{n}}{n} = \frac{1}{\sqrt{n}} \tag{142}$$

goes to zero as a number of trials goes to infinity. So the waiting times between successive change-overs of the winner-looser roles between the contestants are likely to be fantastically long. What should be the pertinent probability density function for such a process? Imagine a huge sample of records of ideal coin-tossing games, each consisting of 2n tosses. We pick one at random and observe the number of the last trial at which the accumulated numbers of heads and tails were equal - that is, the last changeover. This number is even and we denote it by 2K, so that $0 \leq K \leq n$. Frequent changeovers (that is, in the lead of one player) would imply that K is likely to be relatively close to n - but this is not so. Interestingly enough, Feller's calculation gives for the p.d.f. of the variable $x = K/n$ the expression (Fig. 16)

$$P(x) = \frac{1}{n\pi \sqrt{x(1-x)}} \quad . \tag{143}$$

The symmetry of the distribution implies that the inequalities $K \geq n/2$ and $K \leq n/2$ are equally likely. We see that the probabilities near the end points are greatest; the most probable values of K are then the extremes 0 and n, while the median value $K = n/2$ is least likely to occur. The above probability density function is identical (save the proportionality factor 1/n) to the invariant measure of the logistic map $f(x) = 4rx(1 - x)$ for $r = 1$, i.e. for the maximum value of the control parameter where the chaotic regime occupies the whole invariant interval and the map becomes exactly two-to-one.

Let us now turn to the distribution of the occurrence of errors in data transmission on real communication channels, the simplest being the binary symmetric channel (Fig. 17) (Gilbert (1960), Berger and Mandelbrot (1963)). The channel is modelled as a Markov chain with two states G (Good = errorless) and

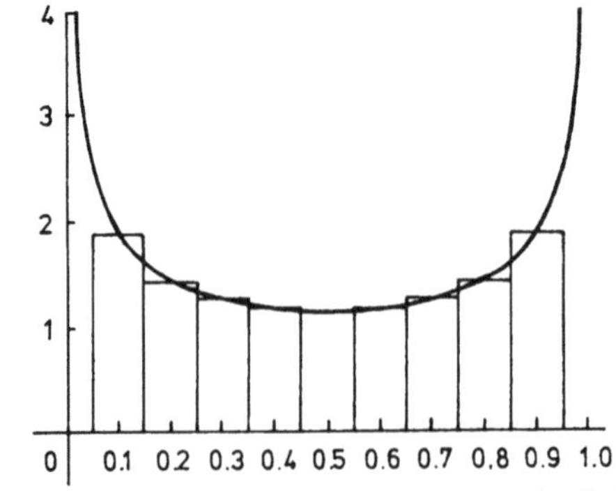

Fig. 16 The probability-density-function of x = k/n (after Feller (1970))

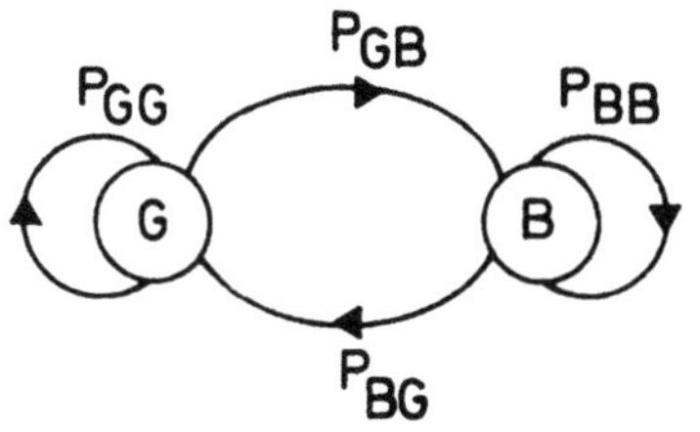

Fig. 17 The Markov-chain model for the burst error channel

B (Bad). The probability of error depends upon the state. State G is associated with the zero probability of an error. In state B a (loaded) coin is tossed to decide whether an error will occur or not. The (biased) coin-tossing feature is included because actual bursts or clusters contain "good" digits interspersed with the errors. To stimulate burst errors, the states G and B must persist; hence the transition probabilities P_{GB} and P_{BG} should be small and the probabilities P_{GG} and P_{BB} should accordingly be large. Berger and Mandelbrot started by considering groups of three successive errors taking the time position t_{n-1} and t_{n+1} fixed and allowing t_n to hover in between; what is the probabilitiy density function of the inter-error intervals and the probability density of t_n? If the inter-error intervals are geometrically distributed (like in some cases assumed by the memoryless binary symmetric channel), the distribution of t_n is uniformly distributed between t_{n-1} and t_{n+1}. Berger and Mandelbrot, in order to bring the model as close as possible to the collected data, assumed a Pareto distribution for the inter-error intervals. In that case, for the probability density of t_n the theoretical type of curve best fitting the experimental results is illustrated in Fig. 18. It says essentially that the probability of the occurrence of an error increases as we approach t_{n-1} or t_{n+1} symmetrically and becomes greatest in the close neighbourhood of the occurrence of another error (the previous or the next). The probability of having an error in bet-

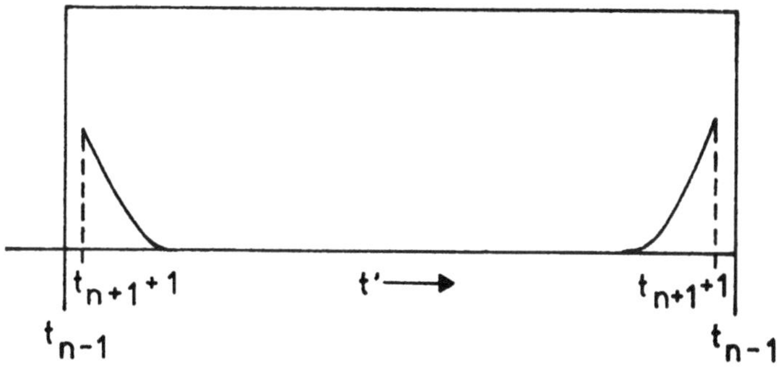

Fig. 18 The p.d.f. of the "intermediate error" in a burst noise channel (after Berger and Mandelbrot (1963))

ween is practically zero. Similarly, in the case of a large number of errors a sizeable portion of the total sample length will be found in a few of the longest error intervals, say L of them, thus creating a pattern in which errors are mostly grouped in L clusters. Again the median value of the distribution is the most improbable one while the greatest probability refers to tight error clustering on time. The shape of the probability density function is again of the "hyperbolic" type we have already encountered twice above. The channel has its "Good" days and its "Bad" days - occurring in the persistent clusters. This persistence and the scarcity of a possible changeover reminds one of the random walk business or the fair tossing again. In fact, one might conjecture that the time axis within the transmission interval displays a fractal property, that is, the clustering is self-similar down to small scales. Most likely the channel spends either too long or too short a time in the errorless regime, the relative number of errors or the average number of digits in errors should be expected to tend to zero as the length of the message n increases to infinity. This conjecture comes for the same reasoning mentioned previously with reference to the average number of changes in the lead of one regime as time outstretches to infinity. It turns out then that, as the length of the transmitted message goes to infinity, despite of the presence of an unbounded number of errors the channel capacity of the burst-error binary symmetric channel tends to one, which is the figure one usually gets for the noiseless memoryless channel. Finally, in their study of intermittent behaviour near a tangent bifurcation of the logistic map, Hirsch, Huberman and Scalapino (1982) derive a similar probability density function for the part lengths that the orbit spends in the laminar (pseudo-limit cycle) regime or the chaotic regime - where it wanders aperiodically before getting a chance of being reinjected in the "channel" formed between the map and the bisector line - that is, in the limit cycle regime. In other words, in the intermittent regime the orbit spends, say within the limit cycle regime, either too little or too much time. Again, the average value of the (symmetrical) distribution function is the least likely to occur.

It appears therefore that in a broad domain of seemingly unconnected physical phenomena a chaotic dynamics or macroscopic "turbulent noise" prevails - in the absence of microscopic multidimensional noise - and in fact it is responsible for the break down of the "law of large numbers".

11. Thermodynamics of (fractal) curves and patterns in the plane

Perception always takes place in two dimensions since information is carried on the surface of spherical waves whose wave number is prefixed - thereby allowing for the carrier only two degrees of freedom (two independent projections of the wave number). It would be highly advisable to have a statistical theory (more concretely, a "Thermodynamics") of chaotic curves and patterns in two dimensions since this would facilitate the building up of a - still not existing - mathematical theory of perception and an unambiguous way of classifying patterns.[1)] Although some comments on this issue advocating against the adoption of the "mean field regime" have already been made in Section 10, here we report on some recent theoretical investigations essentially originating with Mendes-France (1984) (see also references therein).

To begin with, consider first a curve on the plane of finite length Γ and let $\Omega(\Gamma)$ be the set of straight lines D which intersect Γ in any number of points.

The portion of the length of the frontier of the convex hull of Γ (i.e. the length of the envelope around Γ), that is the sum of $d\rho d\theta$ elements associated with the straight lines which cut Γ in exactly k points, equals

$$m_k = \int_{\mathrm{card}(D \cap \Gamma)=k} dD \tag{144}$$

where $dD = d\rho d\theta / \mathrm{meas}(\Omega(\Gamma))$.

So the total length of the envelope of Γ will be

$$\delta K = \sum_{k=1}^{\infty} m_k \quad . \tag{145}$$

The quantity $P_k = m_k / |\delta K|$ can be therefore considered as the probability that the straight line cut Γ in exactly k points. The "entropy" of the curve Γ can be written as

$$S(\Gamma) = - \sum_{k=1}^{\infty} P_k(\Gamma) \ln P_k(\Gamma) \quad . \tag{146}$$

The average number of intersecting points of Γ, N, with a random straight line D is

---

1) As more or less complex.

$$N = \sum_{k=1}^{\infty} KP_k = \frac{\sum_{k=1}^{\infty} Km_k}{|\delta K|} \qquad (147)$$

and after Steinhaus' theorem (1954) it equals $2|\Gamma|/|\delta K|$, hence

$$\sum_{k=1}^{\infty} Km_k = 2|\Gamma|. \qquad (148)$$

The larger the N the more "complex" is the planar shape under study.

Let us try to find the maximum of the function

$$S(\Gamma) = -\sum_{k=1}^{\infty} P_k \ln P_k \qquad (149)$$

subject to the constraints

$$\sum_{k=1}^{\infty} P_k = 1 \quad \text{and} \quad \sum_{k=1}^{\infty} KP_k = 2\frac{|\Gamma|}{|\delta K|}. \qquad (150)$$

After applying the known technique of Lagrangian multipliers we get

$$P_k = \alpha e^{-K\beta}. \qquad (151)$$

The two constraints above give

$$\alpha = e^{\beta} - 1 \qquad (152)$$

and

$$\beta = \ln \frac{2|\Gamma|/|\delta K|}{(2|\Gamma|/|\delta K|) - 1} = \ln \frac{2|\Gamma|}{2|\Gamma| - |\delta K|} \qquad (153)$$

or

$$\beta = \ln \frac{N}{N-1} \qquad (154)$$

respectively. So

$$P_k = (e^{\beta} - 1)e^{-\beta K} \qquad (155)$$

and after some trivial calculations

$$S = \ln \frac{2|\Gamma|}{|\delta K|} + \frac{\beta}{e^{\beta} - 1} = \ln N + \frac{\beta}{e^{\beta} - 1}$$

$$= \ln N + \frac{\ln (N/N-1)}{(N/N-1) - 1} = \ln \frac{N^N}{(N-1)^{N-1}} \tag{156}$$

This is the expression for the entropy of a curve in "equilibrium". We can define by analogy an equivalent "temperature"

$$T \sim \frac{1}{\beta} = \left(\ln \frac{2|\Gamma|}{2|\Gamma| - |\delta k|}\right)^{-1} . \tag{157}$$

It follows immediately that the only curves at $T = 0$ are straight ($|\delta K| = 2|\Gamma|$) line segments. Then $S = 0$. It also follows that <u>any convex closed curve</u> (for such a curve $|\Gamma| = |\delta K|$) has an entropy

$$S = \ln 2 + \frac{\ln 2}{2-1} = 2\ln 2 = \ln 4 . \tag{158}$$

Let us now extend the above formalism to curves of infinite length (fractals for example). Let Γ be a curve of infinite length and let Γ_s represent the beginning portion of Γ of length s. We have seen that the entropy of Γ_s is

$$S(s) = \ln \frac{2s}{|\delta K_s|} + \frac{\beta_s}{e^{\beta_s} - 1} . \tag{159}$$

We normalize this entropy by dividing by a factor $\ln 2s$. We now define the entropy of the curve Γ as

$$S(\Gamma) = \lim_{s \to \infty} \frac{S(s)}{\ln 2s} \tag{160}$$

provided that this limit exists.

Infinite straight lines, exponential spirals ($\rho = \alpha^{\theta}$, $\alpha > 1$) or any curves which go rapidly to infinity have zero entropy. Others, like for example the logarithmic spiral $\rho = \log\theta$, have maximum entropy.

So it happens now that the macroscopic parameter N is invariant under coordinate transformations involving translation, rotation and scale change; so it may serve as an "order parameter" for the pattern concerned. In a recent investigation Shimizu and Tsuda (1983) have tried to pinpoint a neurophysiological analogue for the parameter N, namely the average number of firing cells of the hypercolumn of neurons in the optical cortex. Since the parameter N stands for a frequency, Shimizu and Tsuda (1984) have proposed that instant identification of two patterns can be understood in terms of mutual entrainment of the frequencies N_0, N_1. Briefly, the argument made by Shimizu and Tsuda is that the information contained in a pattern (1) characterized by a parameter N_1 is $I \sim \log N_1$ since the larger value of N_1 indicates a more complex object. On the other hand, if ΔN_0 is the region of entrainment of a template-frequency N_0 in the processor (brain) that is the frequency span within which the oscillator N_1 can be attracted by the oscillator N_0, then the equivocation $I(N_1/N_0)$ equals $-\log(1/\Delta N_0)$ since the stricter the entraining band ΔN_0 the more congenial N_1 appears in contrast to N_0. Hence a mutual information quantity which refers to pattern N_1 obtained by knowledge of N_0 is

$$I(N_1, N_0) = I(N_1) - I(N_1/N_0) = \log(N_1/\Delta N_0) \text{ bits} \quad . \tag{161}$$

12. Concluding remarks

12.1. The syndrome of self-reference

For values of the control parameters below the inition of chaos (for which only steady states and periodic cycles exist) the trajectories in state space will only attract and the corresponding iterations in the Poincaré return map on the interval will monotonically converge. Under such circumstances $\langle I \rangle < 0$, i.e. the system will behave asymptotically as an information sink.

Above the critical value(s) of the control parameter(s) for which continuous stable chaos is established, $\langle I \rangle > 0$ along some variables (information source) and $\langle I \rangle < 0$ along other variables (information compression or dissipation). So much for the information production/dissipation interplay.

A dynamic model of natural language should nevertheless simulate as well as explain a number of key "idioms" such as (a) the ability of the language to form abstractions (collective properties), for instance cross-correlations among underlying stored patterns and environmental messages and (b) the perennial paradox of self-reference.

At first glance one could argue that the syndrome of self-reference might be related to "circular argumentation" which, however, is dynamically attributed to maps folding (iterating) onto themselves; this, in turn, implies the existence only of steady states and periodic orbits giving rise to information dissipation.

The congruence then between "circularity" and self-reference appears rather superficial since circularity implies "halting" of processing, while paradox implies "deadlock" in the cognitive machine.

Indeed the paradox of self-reference goes deeper, namely it has something to do with the inability of the language to discriminate between statements belonging to hierarchically different cognitive levels. (Confusing a set for one of its members signals precisely the appearance of the Russelian paradox.) In short, the paradox of self-reference is associated with the "distress" caused by the inability to compress an irrational number; it can be illustrated in the following linguistic example: "This sentence means what this sentence means".

In any cognitive process the receiver aims at "breaking" the code of the incoming signal, i.e. "compressing" or abstracting as much as possible the description conveyed by the meassage, thereby achieving a high degree of predictability.

Achieving compressibility is therefore tantamount to forming collective properties out of variables of the signal under study or, in the domain of description governed by the solution of the master equation, tantamount to compressing the p.d.f. P(x) to achieve the "mean field" approximation, which means decoupling and preponderance of the first moment(s) of the p.d.f. over higher moments (variance, etc.).

One should recall (Nicolis and Prigogine (1977)) that near bifurcation points, triggering transitions either among steady states, steady states-limit cycles or among periodic orbits and chaos, the p.d.f. P(x) becomes double- or multi-humped (the median ceases to be the most probable value), all moments of the p.d.f. may become comparable and are coupled via the nonlinearity of the transitional probabilities of the master equation. So the "macrodescription" at the level where cognition should take place acquires as many degrees of freedom (moments) as the "microdescription", one hierarchical level below (see Fig. 19).

This is what we mean by saying that in such cases the dynamics on two successive hierarchical levels become, near bifurcation points, indistinguishable.

Therefore, when the control parameters allow only the existence of steady states and periodic orbits, away from bifurcations, the "mean" field regime is ensured, i.e. deliberations at different hierarchical levels - the "microlevel" of variables and the "macrolevel" of moments - are quite distinct. This is ensured by the small spread of the p.d.f. around the mean.

Yet, stagnation results since, in the vicinity of the above two kinds of attractors, information is neither produced nor dissipated. The "machine" simply halts as it happens, for instance, in the calculation of the digits of a rational number. Continuation of the process would literally lead to circularity. If, on the other hand, the control parameters allow for the production of continuous self-sustained chaos, the p.d.f. is broad (it covers the whole interval) and it may also appear quite irregular, depending on the map. This may again signal the breaking of the "mean field" regime. So continuous chaos, in spite of the fact that the strange attractor generates and compresses information at the same time for different variables, is perhaps not the ideal model for a linguistic system.

One should rather settle for an "intermittent" regime where chaos is unstable, so to speak, thereby offering opportunities for spontaneous alternations between the expansion and compression of the p.d.f.

Fortunately, the patterns of the mentioned simple dynamical systems allow for two such distinct behaviours, namely metastable chaos and intermittency.

(a) Metastable chaos: Realized for some windows of the control parameter value, it implies a transitory regime whose decay times have an exponential distribution.

Almost every trajectory eventually settles down to a periodic orbit after some initial period of irregular oscillation. The mean duration of the chaotic regime is not long in general, typically about 50 iterations, but initial conditions can be found by trial and error on the map which remains chaotic much longer.

In other situations a metastable chaotic regime might decay to a steady state or to one which is also chaotic but which has a significantly different behaviour, i.e. different p.d.f. and occupies a different subinterval of the whole attractor or, the new regime might also be metastable, eventually decaying to

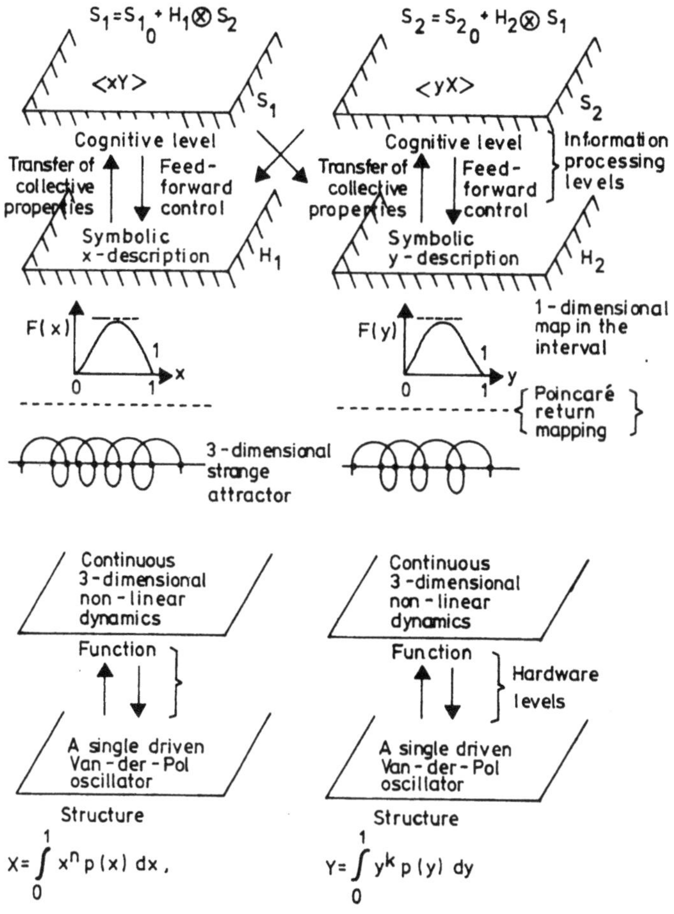

Fig. 19 Layout of the simplest communication scheme between two linguistic, hierarchical systems

a third regime or possibly back to the original one.

(b) Intermittency: Realized for some other windows of the control parameter values of the same dynamical system, it refers basically to the physical inability of achieving perfect entrainment or stable, strict locking between, e.g., the phase of a dissipative non-linear relaxation oscillator and the phase of an external periodic excitation.

By the time the limit cycle in the state space is about to close, a "scrambling" process, as it were, intervenes and destroys the orderly trajectory; for a brief time-interval it creates a chaotic motion amounting to broadband noise and then the building-up of a limit cycle (a different one) starts all over again. After a while a new bursting transition to "turbulent chaos" occurs, followed by another limit cycle appearance and so on, the time sequence of intermittent points being random (they follow a hyperbolic p.d.f.).

To sum up: in the neighbourhood of steady states and periodic orbits the "cognitive machine" halts. Circularity sets in but the p.d.f. is very narrow and the mean field regime works, so abstractions are formed impeccably.

In the vicinity of bifurcations from steady states $\rightleftarrows$ limit cycles $\rightleftarrows$ chaos, the "cognitive machine" suffers a "deadlock"; the p.d.f. explodes, the mean field regime breaks down, and abstractions (i.e. cross-correlations) cannot be formed unambiguously. The system at the level of collective properties is turbulent in the sense that it aquires as many degrees of freedom as at the level of single variables.

In self-sustained stable chaos the situation is not necessarily the same since information is generated along specific directions and compressed along others, still the prospect of a broad and "multispiky" p.d.f. justifies some reluctance in adopting this model for a reliable processor. It seems that the closest model for such a processor should be the regime of intermittency and metastable chaos.

12.2. The origin of conflict in communicating hierarchical systems

In two communicating (symbolic) dynamical systems (Fig. 19) we postulate that conflict, a by-product of communication, emerges as a dilemma of mutually accommodating excessive entropy production (Nicolis and Prigogine (1977)), Nicolis and Protonotarios (1979)). Each system strives towards "self"-organization at the expense of the partner.

The hierarchical system we are dealing with possesses complexity and organization. We define complexity as the minimum number of bits required to reproduce a given system. We define organization as the ability to compress information, which in turn is generated via cascading bifurcations giving rise to broken symmetry or, (beyond chaos) via cascading iterations on the map in the interval. "Self"-organization is meant to imply here the function of a cognitive gadget, language, which compresses the complexity of the "opponent" (environment) thereby providing minimal algorithms which reduce and predict the behaviour of the partner system.

In any "game" of this sort each participant therefore aims at "breaking" the code of the opponent, i.e. "compressing" as mush as possible the description of the partner. Achieving compressibility (or cognition) is therefore tantamount to forming collective properties out of the variables of the system under study or, in the domain of a macrodescription governed by the hierarchy of the moments of the master equation, tantamount to compressing the p.d.f. to achieve the "mean" field regime (see below).

Usually, languages are recursive hierarchical algorithms (see Fig. 19). There are cases, however, where languages may (a) become involved in circularity (which means "halting" from the point of view of a processing function) or (b) come to a "deadlock" which may signal the appearance of what is known as the paradox of self-reference. Linguistic paradoxes emerge whenever the attempt to compress the time series received from the environment and which effort produces excessive entropy production at the hardware level of the processor involved reaches the so-called "Gödelian" limit; beyond this limit the incompatibility between completeness and self-consistency of any closed axiomatic (logical) system manifests itself as an inability to discriminate between statements (strings of symbols) belonging to different hierarchical levels.

When paradoxes flare up within a language-possessing system, conflict sharpens; to the previous inter-systemic conflict we now witness a hierarchically underlying intra-systemic conflict.

Inter-systemic conflict resolution depends now on the parameters determining the intra-systemic conflict. The latter, in turn, uses the inter-systemic communication as an expendable device for advancing its own resolution. The coupling and dynamic evolution of the pair of inter- and intra-systemic conflict were recently discussed (Nicolis, Protonotarios and Theologou (1978), Nicolis and Protonotarios (1979, 1980)) in detail.

Dynamically speaking, we may say that the origin of conflict can be boiled to the incompatibility between compressibility and intermittency; if the p.d.f. cannot be compressed beyond a certain threshold and explodes spontaneously ("intermittency"), the mean-field approach breaks down and the system processor at the level of moments becomes "turbulent" in the sense that it aquires as many degrees of freedom as the system at the level of bare variables (one hierarchical level below).

This is what is meant by saying that beyond the "Gödelian" limit the dynamics of two successive hierarchical levels get "mixed up", i.e. microscopic and macroscopic descriptions from the point of view of time series become indistinguishable in the sense that random noise and chaos are statistically similar (or they appear to be, so far).

In many instances hierarchical systems are "almost decomposable" in that one can study the dynamical ongoings at a given level, taking what is just underneath as a boundary condition and what takes place above as a constant. This can be done whenever the rate constants differ by orders of magnitude from level to level (lower level being characterized by larger rate constants or smaller relaxation times).

In linguistic systems, however, this decoupling of levels cannot be accomplished due to the existence of evolutionary feed forward loops mediating the dynamics between successive levels. Specifically, in linguistic systems, one makes use, at the same time, of two hierarchical levels where statements and meta-statements are interlocked. ("Chicago" is a trisyllabic word and a city.)

Let us now examine the cirmustances of breaking down the "mean-field approximation" in more detail. Consider a physical dynamical system reduced to a one-dimensional map. The probability of finding the system at point x at time t increases due to transition from other points x' of the interval. It decreases due to transitions leaving this point, i.e.

$$\frac{d}{dt} P(x; t) = \text{rate in} - \text{rate out} = I - I' \quad . \tag{162}$$

Since the term I consists of all transitions from initial points x' to x, it is composed of the sum over the initial points. Each term is given by the probability to find the system at a point x' multiplied by the transition probability per unit time to pass from x' to x. Thus

$$I = \sum_{x'} w(x, x')P(x'; t) \quad , \tag{163}$$

where $w(x, x')$ is the transitional probability for jumping from $x' \rightarrow x$.

For the outgoing transitions I', we have

$$I' = P(x; t) \sum_{x' \neq x} w(x', x) \quad , \tag{164}$$

where $w(x', x)$ is the transitional probability for jumping from $x \rightarrow x'$. So, the master equation becomes

$$\frac{dP(x; t)}{dt} = \sum_{x'} w(x, x')P(x'; t) - P(x; t) \sum_{x' \neq x} w(x', x) \quad . \tag{165}$$

The "steady state" or "stationary" asymptotic solution of this equation is what was investigated in Eq. (106). Now the crucial thing is that the transitional probability rates (generally unknown, but depending on the shape of the map) are in most cases non-linear functions of x. In chemical reactions, for instance, these rates are, according to the law of mass action, proportional to the concentrations of the "reacting species", in other words, to the number of ways a pair or a group of "reactant molecules" can be sorted out from the total population.

For example, for the reaction $2x \rightarrow x'$ the rate is proportional to $(x/2)(x-1)$, i.e. the equation for the "mean" is

$$\frac{d \langle x \rangle}{dt} = -K \langle x(x - 1) \rangle \quad . \tag{166}$$

Multiplying now both members of (165) by x, x^2, ..., and integrating or summing in the interval of x we get a series of "phenomenological", macroscopic equations with reference to the various moments $\langle x \rangle$, $\langle \delta x^2 \rangle$, ..., of the p.d.f. $P(x, t)$ namely,

$$\begin{aligned} \frac{d}{dt} \langle x \rangle &= f_1 \{ \langle x \rangle , \langle \delta x^2 \rangle , \ldots \} \\ \frac{d}{dt} \langle \delta x^2 \rangle &= f_2 \{ \langle x \rangle , \langle \delta x^2 \rangle , \ldots \} \\ &\vdots \qquad\qquad \vdots \end{aligned} \tag{167}$$

Because of the non-linearity of the operators f_i (due to the non-linearity of the transitional probabilities) we get a great number (infinite) of coupled non-linear differential equations with respect to the moments and cross-moments (cross-correlations). Moreover, in the vicinity of bifurcations, or, for self-sustained chaos filling the whole interval, these moments may become comparable.

We then understand very clearly that under such circumstances successive hierarchical levels become indistinguishable, a necessary prerequisite for the appearance of the paradox of self-reference and the "dead-lock" of a linguistic processor.

12.3. Global (analogous, synchronous) versus local (digital, sequential) pattern recognition

In the Introduction and in Section 11 of this paper we investigated criteria for the complexity of a time series and the complexity of a (fractal) continuous curve, respectively.

In the case of a time series, complexity is the length in bits of the shortest program needed to reproduce the pattern. In the case of a continuous curve, complexity can be associated with a global "order parameter" N which stands for the average number (or the frequency) of intersections of the curve by a set of straight lines. This order parameter may allow for spontaneous cognition of the object via the well-known mechanism of entrainment or phase locking between the incoming frequency and an attractor-template basic frequency within the processor. As communication engineers have known for a long time this is exactly the way an analogous phase-locked loop (P.L.L.) works. Most probably a biological processor uses both modes of cognition. Frequency fusion, phase locking or, more generally, classification of incoming stimuli in a few coexisting attractors allows for the swift creation of a (coarse) "background".

Sequential or intermittent scanning (via saccadic eye movements for example - see Section 10) on the other hand allows the progressive fixation of the details in an already established "context" via cascading iterations. Chaotic dynamics, as we have proposed in this paper, may mediate both modes of cognition.

Finally, the question "where exactly information is stored?" seems to us to receive a rather original answer from a series of well-known experiments on the perception of random dot interference patterns performed several years ago by Glass and Perez (1973).

The essential conclusion we draw from those experiments is that the information lies neither in the externally impinging pattern (which may be just a random dot aggregation on the plane) nor in the dynamically stored pattern - replica of the receiver (which again may by random); it rather emerges from the flow in the neighbourhood of the singularities - as deduced from the pertinent eigenvalues of the linear or non-linear, discrete or continuous transformation (mapping or flow) - which transformations the processor imposes between the incoming and the internal pattern.

So one is inclined to consider, at least in a number of cases where such experimental documentation exists, that what we "really" see or cognize in the world are perhaps just the attractors (or the repellors) of the maps or the flows which the processor generates by way of relating external stimuli with internal activity. Biological processors belonging to the same species possess identical nervous systems and they accordingly tend to generate more or less identical maps or flows - thereby creating universal pattern classification schemes by "social consensus".

Not only in perception but also in Cognition and Linguistics, emerging stable patterns may be identified as eigenvalues of non-linear operators which are involved in the recursive character of these systems. For example, recent work in game theory as applied to biology and animal contests (J. Maynard-Smith (1982)) as well as to the process of mutual simulation between competing hierarchical systems - each trying to "break" the code of the opponent first, (J.S. Nicolis et al. (1978, 1979)) - shows that rationality in such tournaments can be completely replaced by evolutionary stability. This points to the emergence of evolutionary stable strategies (E.S.S), which are point attractors for the probabilities of moves adopted by the contestants: Such asymptotically stable strategies lead to a stable population polymorphism or stable behavioural repertoires adopted by a single individual.

In Linguistics the situation is different since, here, unlike animal or finite state automata games, self-reference is involved. We conclude with two examples which show that self-consistency, hence logic, is provided also by the existence of stable attractors.

Take as a first example the symbolic self-referential loop:

"This sentence contains $\underline{X_0}$ 0's, $\underline{X_1}$ 1's, $\underline{X_2}$ 2's, $\underline{X_3}$ 3's, $\underline{X_4}$ 4's, $\underline{X_5}$ 5's, $\underline{X_6}$ 6's, $\underline{X_7}$ 7's, $\underline{X_8}$ 8's, $\underline{X_9}$ 9's".

Let the 10-dimensional vector $\underline{X}$ $(X_0, X_1, X_2, \ldots, X_9)$ start from a randomly selected initial condition and then provide the next iteration by checking the truth of the previous one.

For example start from $\underline{X}_a$ (0,1,2,3,4,5,6,7,8,9).

The next iteration gives $\underline{X}_b$ (2,2,2,2,2,2,2,2,2,2)

Next comes: $\underline{X}_c$ (1,1,11,1,1,1,1,1,1,1), followed

by $\underline{X}_d$ (1,12,1,1,1,1,1,1,1,1) and finally

$\underline{X}_e$ (1,11,2,1,1,1,1,1,1,1) , which obviously

is the attractor corresponding to the selected initial condition and which makes the self-referential sentence above self-consistent, that is, true. (Through exhaustive search one can find another point attractor as well as a limit cycle of period 2. Here the basin of attraction is partitioned into three subsets.)

As a second example take the sentence:

"This sentence has $\underline{X}$ letters" .

Start iterating from X_a = "one" .

Then $F(X_a)$ = 25, where F is the number of letters after putting the given initial condition into the sentence.

The second "seed" then is X_b = "twenty five", which leads to $F(X_b)$ = 32. Likewise X_c = "thirty two" leading to $F(X_c)$ = 31 and finally X_d "thirty one" leads the system to $F(X_d) = 31$ - which is the attractor corresponding to the initially chosen seed "one". (there is another point attractor corresponding to a different subset of the basin.)

Note finally that, while the truth assessed by the second sentence is "English bound" (translated e.g. into French it becomes false), the first sentence conveys a "translinguistic" truth.

Appendix: A physical system composed of N hard spheres

Let us consider in detail a physical system composed of N hard spheres. This system of moving scatterers is sensitive to initial conditions in the following sense:

Take the collision process between a pair of spheres. An angular deviation θ on the trajectory of a sphere changes after a collision to an angle θ' where $\theta' \sim \frac{2\delta}{R}\,\theta$, where δ is the mean free path and R is the diameter of each sphere. Let τ be the intercollisional time (which is of the order of 10^{-10} sec). After a time t the deviation of the trajectory of one sphere in the 6N-dimensional state space will grow as $\left(\frac{2\delta}{R}\right)^{t/\tau}$, so the maximum (positive) Lyapunov exponent of the system is $\lambda \sim \frac{1}{\tau} \log\frac{2\delta}{R}$.

Since the Kolmogorov-Sinai entropy is taken equal to the sum of positive Lyapunov exponents and since in our system <u>all</u> Lyapunov exponents are positive, the K-S entropy will be $S \sim N \cdot \lambda$, where N is of the order of the Avogadro number. Taking into account a typical varified "Lorentz gas" we get: $S \sim 10^{34}\,\frac{\text{bits}}{\text{sec}}$ - which represents essentially the <u>new</u> information (or variety) <u>generated</u> by the evolving system per time unit - or the information needed to follow the second-to-second trajectory of the system.

References

Adey, W.R., 1968: in The Mind: Biological Approaches to its Function, ed. by Corning, Balaban (Wiley, New York), pp. 69-99

——, 1970: The Neurosciences: A Study Program, ed. by F.O. Schmitt (Rockefeller University Press, New York), p. 181

——; 1972: Int. J. Neurosci. **3**, p. 271

Anderson, P.W., 1972: Science **177**, p. 393

Benettin, G., 1983: "Mean Field Approximation for Lyapunov Exponents in Some Conservative Dynamical Systems", Preprint

Benettin, G., L. Galgani, A. Giorgilli and J.M. Strelcyn, 1978: C.R. Acad. Sci. Paris **286**, A-431

Berger, J.M. and B. Mandelbrot, 1963: IBM Journal (July), p. 224

Bremermann, H.J., 1982: Int.J. Theor.Phys. **21**, p. 203

Chaitin, G., 1975: Sci.Am. (May), p.47

Davies, P.C.W., 1977: The Physics of Time Asymmetry (University of California Press)

Elul, R., 1966a: Nature, Lond. **210**, p. 1127

——, 1966b: "Statistical Mechanisms in Generation of the EEG", in Progress in Biomedical Engineering, ed. by L. Fogel (Spartan Books, Washinton, D.C.), pp. 131 - 150

——, 1968: "Brain Waves: Intracellular Recording and Statistical Analysis Help Clarify their Physiological Significance", in Data Acquisition and Processing in Biology and Medicine, ed. by Enslein (Pergamon, Oxford), pp. 93 - 114

——, 1969a: "Scanning of Cortical Neurons in the EEG: A Possible Mechanism of Attention and Consciousness", in Information Processing in Dendrites, ed. by A. Scheibel, M. Scheibel (Brain Information Service, University of California, Los Angeles)

——, 1969b: Science **164**, p. 328

——, 1972: "Randomness and Synchrony in the Generation of EEG", in Synchronization of EEG Activities in Epilepsies, ed. by H. Petsche, M. Brazier (Springer, Berlin, Heidelberg, New York), pp. 59 - 77

Farmer, J.D., 1982: Z.Naturforsch. **37a**, p. 1304

Farmer, J.D., E. Ott and J.A. Yorke, 1983: Physica **7D**, p. 153

Feigenbaum, M.J., 1980: Los Alamos Science (Summer), p. 4

Feistel, R., 1983: "Language and Information Content", Preprint

Feller, W., 1970: An Introduction to Probability Theory and Applications, Third Ed., Vol. 1 (Wiley, New York)

Gilbert, E.N., 1960: The Bell Syst. Techn. J. (September), p. 1253

Glass, L. and R. Perez, 1973: "Perception of Random Dot Interference Patterns", Nature **246**, (December 7), pp. 360 - 362

Grassberger, P. and I. Procaccia, 1984: Physica 13D, p. 34
Grebogi, C., E. Ott and J. Yorke, 1983a: Phys. Rev. Lett. 50, pp. 935 - 938
Grebogi, C., S. McDonald, E. Ott and J. Yorke, 1983b: Phys. Lett. 99A
Haken, H., 1978: Synergetics: An Introduction, 2nd Ed. (Springer, Berlin, Heidelberg, New York)
——, 1983: Phys. Lett. 94A, p. 2
——, 1984: Advanced Synergetics (Springer, Berlin, Heidelberg, New York)
Hirsch, J.E., B.A. Huberman and D.J. Scalapino, 1982: Phys. Rev. A25, p.519
Kilmer, W. and W.S. McCulloch, 1969: "The Reticular Formation Command and Control System", in The Information Processing in the Nervous System, ed. by Leibovic (Springer, Berlin, Heidelberg, New York)
Lichtenberg, A.J. and M.A. Liebermann, 1983: Regular and Stochastic Motion (Springer, Berlin, Heidelberg, New York)
Longuet-Higgins, H.C., 1968: "The Non-Local Storage of Temporal Information", Proc. Roy. Soc. B171, pp. 327 - 334
Lorenz, E.N., 1964: Tellus 16, p. 1
Mandelbrot, B., 1982: The Fractal Geometry of Nature (Freeman, New York)
Manneville, P. and Y. Pomeau, 1980: Physica 1D, p. 219
Mason, S.F., 1984: "Origin of Biomolecular Handedness", Nature 311, (Sept.), pp. 19 - 23
May, R.M., 1979: Int.Conf. on Non-Linear Dynamics, New York (Dec. 17 - 21); 1981: Ann. N.Y. Ac. Sci. 357
Mendes-France, M., 1984: "Folding Paper and Thermodynamics", Phys. Reports 103, No. 1-4. pp. 161 - 172
Miller, G.A., 1974: The Psychology of Communication (Penguin)
Mori, H., 1980: Progr. Theor. Phys. 63, p. 1044
Moruzzi, G. and H.W. Magoun, 1949: Clin. Neurophys. 1, pp. 455 - 473
Nicolis, G. and I. Prigogine, 1977: Self-Organization in Non-Equlibrium Systems (Wiley, New York)
——, 1981: "Symmetry Breaking and Pattern Selection in Far-From-Equilibrium Systems", Proc. Nat. Ac.Sci. USA 78, No. 2, pp. 659 - 663
Nicolis, J.S., 1982a: "Sketch for a Dynamical Theory of Language", Kybernetes 11, (April), pp. 123 - 132
——, 1982b: "Should a Reliable Information Processor Be Chaotic?", Kybernetes 11, (October), pp. 269 - 274
——, 1984: "The Role of Chaos in Reliable Information Processing", J. of the Franklin Inst. 317, No. 5, (May), pp. 289 - 307; also in Synergetics, ed. by H.Haken, Proceedings of the Elmau Conference, 1 - 8 May 1983.
——, 1985: Dynamics of Hierarchical Systems. An Evolutionary Approach, Vol. 25 (series of Synergetics ed. by H.Haken), (Springer, Berlin, Heidelberg, New York)

Nicolis, J.S. and M. Benrubi, 1976: J. Theor. Biol. 59, pp. 76 - 96
Nicolis, J.S., G. Nicolis and M. Benrubi, 1977: In Proc. 1st Int. Conf. Infor. Sci. Systems, ed. by N.S. Tzannes and D. Lainiotis, Vol. 2, (Hemisphere Pub. Co., Washington, D.C., London), p. 569
Nicolis, J.S., E.N. Protonotarios and M. Theologou, 1978: Int. J. Man-Machine Studies, Vol. 10, p. 343
Nicolis, J.S. and E.N. Protonotarios, 1979: Int. J. Biomed. Computing, Vol. 10, pp. 417 - 447
Nicolis, J.S. and E.N. Protonotarios, 1980: Int. J. Syst. Sci., Vol. 11, No. 1, pp. 107 - 127
Nicolis, J.S., G. Mayer-Kress and G. Haubs, 1983: Z. Naturforsch. 36a, pp. 1157 - 1169
Nicolis, J.S. and I. Tsuda, 1985: "Chaotic Dynamics of Information Processing: "The 'Magic Number Seven Plus-Minus Two' Revisited", Bull. Math. Biol. (in press)
Noton, D., 1970: IEEE Trans. on Systems, Man and Cybernetics, SSC-6 4, p. 349
Noton, D. and L. Stark, 1971a: Sci. Amer. 224, (June), p. 34
——, 1971b: Science 171, p. 308
Ott, E., W.D. Witters and J.A. Yorke, 1984: "Is the Dimension of Chaotic Attractors Invariant Under Coordinate Changes?", Journ. of Statist. Physics 36, pp. 687 - 697
Rappoport, A., 1956: "The Promise and Pitfalls of Information Theory", Behavioral Sci. 1, pp. 303 - 315
Schalkwijk, J.P.M., 1982: "On the Quantitative Definition of Information and its Impact on the Field of Communications", Rep. Prog. Phys. Vol. 45, pp. 1213 - 1260
Shaw, R.S., 1981: Z. Naturforsch. 36a, p. 80
Shimada, I. and T. Nagashima, 1979: Progr. Theor. Phys. 61, p. 6
Shimizu, H. and I. Tsuda, 1983: Preprint
Steinhaus, H., 1954: "Length Shape and Area", Colloq. Math. 3, pp. 1 - 13
Takeyama, K., 1978: "Dynamics of the Lorenz Model of Connective Instabilities", Prog. Theor. Phys. 60, pp. 613 - 615
Verzeano, M., 1972: "Pacemakers, Synchronization and Epilepsy", in Synchronization of EEG Activity in Epilepsies, ed. by H. Petsche, M. Brazier (Springer, Berlin, Heidelberg, New York), p. 154
Verzeano, M. et al., 1970: "Evoked Responses and Network Dynamics", in The Neural Control of Behavior, ed. by R.E. Whalen, R.F. Thoson, M. Verzeano and S. Weinberger (Academic Press, New York)

Additional reference

Babloyantz, A. and C. Nicolis, 1985: Preprint, Université Libre de Bruxelles
Maynard-Smith, J., 1982: Evolution and the Theory of Games (Cambridge Univ. Press)

Bei Fragen zur Produktsicherheit wenden Sie sich bitte an:
If you have any questions regarding product safety,
please contact:

Walter de Gruyter GmbH
Genthiner Straße 13
10785 Berlin
productsafety@degruyterbrill.com